"Most of the Good Stuff"

Memories Of

RICHARD FEYNMAN

"Most of the Good Stuff"

Memories Of

RICHARD FEYNMAN

Editors:
LAURIE M. BROWN · JOHN S. RIGDEN

AIP

American Institute of Physics
New York

American Institute of Physics

335 East 45th Street

New York, NY 10017-3483

Library of Congress Cataloging-in-Publication Data

Most of the Good Stuff/editors, Laurie M. Brown, John Rigden
    p. cm.
  Includes bibliographical references.
  ISBN 0-88318-870-8
    1. Feynman, Richard Phillips. 2. Physics—History.
3. Physicists—United States—Biography. I. Feynman, Richard Phillips.
II. Brown, Laurie M.   III. Ridgen, John S.
   QC16.F49A3   1993
   530'.092—dc20   92-46471
    [B]   CIP

TABLE OF CONTENTS

The Research Physicist at Cal Tech

The Teacher at Cal Tech

The Public Physicist and Consultant

Feynman-The Man

INTRODUCTION

Laurie M. Brown and John S. Rigden

Richard Phillips Feynman (whom many people called Dick) was born on May 11, 1918 in the borough of Queens in New York City. He died on February 15, 1988 in Altadena, near Pasadena, where he was Professor of Physics at the California Institute of Technology. Dick had struggled courageously (one could even say, heroically) against his final illness, cancer, undergoing extensive surgery four times over ten years without ever losing the youthful drive and brilliance for which he was famous. In a front-page obituary in the New York Times, James Gleick quoted some extraordinary tributes to him: Sidney Drell called Feynman "the most creative theoretical physicist of his time and a true genius." Freeman Dyson said that he was "the most original mind of his generation." Hans Bethe, borrowing a phrase from the mathematician Mark Kac, said that Feynman was not an ordinary genius but a magician, that is, one "who does things that nobody else could ever do and that seem completely unexpected."[1]

These characterizations by physicists who knew him well contrast sharply with, and help to balance, the other public image that Dick himself loved to project. His New York accent was always present, and in those situations where refined and cultivated syntax was the means of discourse—especially when the urbane words masked an underlying ignorance—Feynman's accent could become pronounced to the point of burlesque. He lightly bore such honors as the Einstein Medal

and the Nobel Prize, yet he admittedly enjoyed the status of "celebrity."
In his best-selling book, **Surely You're Joking, Mr. Feynman**, he casts
himself as the playful showman, who is at first undervalued, because
of his rough manners.[2] In the end, of course, he triumphs through
native cleverness, psychological insight, common sense, and the famous Feynman humor.

Which portrayal is more accurate: the magician of science or the
showman? The answer is that both are accurate. We stand in awe of
the wizardry he exhibited in his physics and we take delight in his
vaudevillian personality. At the same time, these characterizations
leave out a great deal: insatiable curiosity, deep love of nature, a passion for teaching, and high standards of scientific integrity. His interests extended to ants and paramecia; to hypnotism and the mixing of
paints; to spinning dinner plates, locks, codes, and ancient scripts.
From his parents he learned to value and befriend nature, and he tried
to communicate this strongly felt emotion through his teaching, and in
later years also through painting. He was a classroom virtuoso, who
judged his own understanding of the most subtle concepts by his ability to explain it to a novice. Underlying all of these attributes was his
own love of physics. Whatever else Dick Feynman may have joked
about, his love for physics approached reverence.

Sylvan S. Schweber has pointed out that the claim to a "powerful
sense of social irresponsibility," which Feynman said he had learned at
Los Alamos from John von Neumann, was somewhat exaggerated.[3] For
example, at Cornell, following his Los Alamos experience, he gave public lectures on the problems of atomic weapons and atomic power.
Later, he was much concerned about and contributed to the science
education of young people, as well as the general public. And his important role in the investigation of the *Challenger* disaster made him a
well-known public figure.[4]

This book is organized around the impact that Feynman made on
the world of physics through his scientific work, his teaching, and his
personal characteristics. The contributors include his doctoral father,
John Wheeler, his sister Joan Feynman, who is also a physicist, and his
co-workers, collaborators, competitors—those who know his scientific
achievements best. Inevitably, they all tell anecdotes that enlighten and
amuse; however, their aim is not primarily to amuse, or to offer re-
dundant praise, but to recall what Dick Feynman meant and means to
physicists. There is an important sense in which all modern physicists
are Feynman's students.

Most of the articles in the book were presented on January 21, 1988 at an all-day joint session of three scientific societies (in itself an unusual tribute). These societies, meeting simultaneously in San Francisco, were: The American Association for the Advancement of Science, The American Physical Society, and the American Association of Physics Teachers. The morning session, chaired by Laurie Brown, included talks by John Wheeler, Freeman Dyson, Julian Schwinger, and Murray Gell-Mann. At the afternoon session, chaired by Eugen Merzbacher, the speakers were James Bjorken, David Pines, David Goodstein, and Daniel Hillis. These talks, plus an additional short piece by Valentine Telegdi, were published in a special Richard Feynman Memorial Issue of *Physics Today* in February 1989 (another exceptional tribute, thanks to Gloria Lubkin, the Editor of *Physics Today*). The present work contains the papers mentioned, as well as additional reminiscences of Feynman by Hans Bethe, Marvin Goldberger, Michael Cohen, Laurie Brown, John Rigden, and Joan Feynman, to whom we are also grateful for the priceless early photographs.

To begin this appreciation of Feynman's work, consider *The Feynman Lectures on Physics*.[5] These published lectures formed the core of a two-year "elementary" physics course that Feynman gave at Caltech. It is generally agreed that while they are too advanced for most beginning students, they are marvelously instructive and stimulating for teachers. The Feynman Lectures are a good place to discover the intellectual excitement that Dick brought to physics. Most of that spirit remains in the published lectures, even though they have been extensively edited and refined by his collaborators.

The first three chapters give an account of Feynman's approach to science, being an outline of basic physics and its relation to the other sciences, as well as a statement of how science differs from other studies. Thus, abstracting a few sentences (with original emphasis used throughout):

> The principle of science, the definition, almost, is the following: *The test of all knowledge is experiment.* Experiment is the *sole judge* of scientific "truth." But what is the source of knowledge? Where do the laws that are to be tested come from? Experiment, itself, helps to produce these laws, in the sense that it gives us hints. But what is also needed is *imagination* to create from these hints the great generalizations—to guess the wonderful, simple, but very strange patterns beneath them all, and then to experiment to check again whether we have made the right

guess. This imagining process is so difficult that there is a division of labor in physics: there are *theoretical* physicists who imagine, deduce, and guess at new laws, but do not experiment; and then there are *experimental* physicists who experiment, imagine, deduce, and guess.[6]

Notice at the end the sly suggestion that the theorist is an experimenter manqué!

Next Feynman points out that experiments are at best only approximate, and so "we first find the 'wrong' [laws], and then we find the 'right' ones." As an example, he cites the law of constancy of mass with motion. For low speeds our law is a good approximation, but for high speeds it is incorrect, "and the higher the speed, the more wrong we are." But:

Finally, and most interesting, *philosophically we are completely wrong* with the approximate law. Our entire picture of the world has to be altered even though the mass changes only by a little bit [i.e., for low velocities]. This is a very peculiar thing about the philosophy, or the ideas, behind the laws. Even a very small effect sometimes requires profound changes in our ideas.[7]

This passage is quoted here to show that Feynman, far from rejecting philosophy, as he liked to claim he did, was himself a philosopher. A similar statement can be made about Feynman and mathematics, whose rigor he pretended to disdain.

Feynman, like most physicists, adhered to a reductionist philosophy, although some physicists are reluctant to admit it. Not Dick!:

Everything is made of atoms. That is the key hypothesis. The most important hypothesis in all biology, for example, is that *everything that animals do, atoms do.* In other words, *there is nothing that living things do that cannot be understood from the point of view that they are made of atoms acting according to the laws of physics.*[8]

But even Feynman had to mitigate the awful implications of this bold statement, and he concluded the discussion with this: "When we say we are a pile of atoms, we do not mean that we are *merely* a pile of atoms, because a pile of atoms which is not repeated from one to the other might well have the possibilities which you see before you in the mirror."[9]

In approaching the scientific method, he said in the lectures, "We shall have to limit ourselves to a bare description of our basic view of what is sometimes called *fundamental physics*, or fundamental ideas which have arisen from the application of the scientific method." And he continued:

> What do we mean by "understanding" something? We can imagine that this complicated array of moving things which constitutes the "world" is something like a great chess game being played by the gods, and we are observers of the game. We do not know what the rules of the game are; all we are allowed to do is to *watch* the playing. Of course, if we watch long enough, we may eventually catch on to a few of the rules. *The rules of the game* are what we mean by *fundamental physics*...If we know the rules, we consider that we "understand" the world.[10]

As a final selection from the *Lectures*, we consider the comparison of physics with the other natural sciences:

> There is another *kind* of problem in the sister sciences which does not exist in physics; we might call it, for lack of a better term, the historical question...There is no historical question being studied in physics at the present time. We do not have a question, "Here are the laws of physics, how did they get that way?" We do not imagine, at the moment, that the laws of physics are somehow changing with time, that they were different in the past than they are at present. Of course they *may* be, and the moment we find they *are*, the historical question of physics will be wrapped up with the rest of the history of the universe, and then the physicist will be talking about the same problems as astronomers, geologists, and biologists.[11]

From these few excerpts, we can see that Feynman's philosophy of science amounts to what some philosophers have referred to as the standard "scientist's account," by which they usually imply substantial ignorance of or indifference to the issues considered by philosophers, and to some extent by historians, of science. We can see, however, that while Feynman generalized about other savants, e.g., poets, philosophers, and social scientists, in a way which often sounded anti-intellectual, he *was* willing to consider the possibility of a radical reformulation of what he meant by *fundamental physics*, but only if forced to do so by experimental evidence.

Turning from Feynman's explanations and advice to the youth (and to their teachers), let us consider some of his contributions to physics itself, and to its ethos. Our selection is incomplete, personal, and we surely leave out some of other physicists' favorite works. We would like to emphasize those we think are most original and, at the same time, contributed greatly to Feynman's own development, as well as to the ideas of physics. In our opinion, they are the following: the Wheeler–Feynman absorber theory of classical electrodynamics; the path integral formulation of quantum mechanics, quantum fields, and quantum statistics; Feynman diagrams; and the regularization and renormalization of quantum field theories, including quantum electrodynamics (QED).

Some other major contributions that we shall not consider are: the theory of polarons; the theory of superfluidity; quantum gravity; the $V-A$ theory of weak interactions and the conserved vector current (with Gell-Mann); the theory of quark partons. Many theorists would be proud to have written other papers of Feynman that we have not included in *either* of these two lists.

The work that Feynman did during the decade, after he entered graduate school at Princeton in 1939 and began his association with John Wheeler, was all interconnected; it led not only to the achievements for which he shared the 1966 Nobel Prize in Physics, as he explains in his Nobel lecture,[12] but greatly influenced nearly all his subsequent research in physics. It is impossible to improve on the lively and personal account that Feynman gave about his approach to QED in his unusual Nobel lecture, so we will refer to it and quote it when appropriate.

The Wheeler-Feynman electrodynamics was an attempt to reformulate classical electrodynamics (CED) in such a way that its quantum mechanical version, QED, would be free of the difficulties that made it logically inconsistent. Perhaps even more importantly, QED was inapplicable at high energy, as well as at low energy when, in the latter case high precision was required. The difficulties took the form of predictions of infinity for quantities known to be finite, including the mass and the charge of the electron. These so-called divergences could not be ignored, as they led to predictions of infinity for other observable quantities, such as energy shifts, if calculated in any approximation other than the lowest. The infinite mass for a point electron came about, in both CED and QED, because of the action of the electron's own field on itself. According to Feynman's Nobel lecture, he began with the idea (as an undergraduate at MIT!) that this problem could be solved simply by eliminating the concept of the field:

> Well, it seemed to me quite evident that the idea that a particle acts on itself...is not a necessary one—it is a sort of silly one, as a matter of fact. And so I suggested to myself that electrons cannot act on themselves; they can only act on other electrons. That means there is no field at all... There was a direct interaction between charges, albeit with a delay.[13]

And he goes on to say: "That was the beginning and the idea seemed so obvious to me and so elegant that I fell deeply in love with it."

On arriving at graduate school in Princeton, Feynman brought his idea to Professor John Wheeler—and discovered what was wrong with it: Along with self-interaction, Feynman's proposal also seemed to eliminate radiative reaction. Thus a hot object would radiate without losing energy. It was to restore this necessary feature of CED that Wheeler and Feynman together made a new theory that employed advanced as well as retarded solutions of Maxwell's equations and also took into account explicitly the presence of absorbing material (essentially the universe itself).[14] Using potentials which were half-advanced and half-retarded (but which were shown to produce no *observable* advanced effects) made it possible to formulate the basic laws of classical physics, except for gravitation, in the form of a relativistic principle of least action. In their article of 1949, the authors quote their version of this principle (with acknowledgments to K. Schwarzschild, H. Tetrode, and A. D. Fokker), together with the statement: "All of mechanics and electrodynamics is contained in this single variational principle."[15]

In his Nobel lecture, Feynman stressed the overall space-time view of the Wheeler-Feynman work, which Dyson recognized as related to Heisenberg's relativistic S matrix. (Heisenberg's theory was developed during the wartime, and so it was not known to the Princeton researchers). Feynman contrasted it with the Hamiltonian formulation, according to which one views the unrolling of space–time like a scroll of which we see only one "plane" constituting all of space at a given time (with appropriate modification for relativity). Eventually, all correct descriptions of nature are found to be equivalent, but Feynman took care to point out—in fact, that *is* the main message of his Nobel lecture—that each different viewpoint can suggest different interesting generalizations, as well as providing new calculational methods. This is an exhortation to avoid faddishness in science.

An important example of a new insight gained from the overall space-time view is the "backward-in-time" description of positrons. Another is the invention of Feynman graphs, which serve as calculational aids, and also as a tool for reasoning. Feynman graphs have had profound applications in virtually every aspect of quantum physics. Although the algorithms embodied in these diagrammatic methods are often referred to as the "visualization" of processes of creation, annihilation, and scattering, this must be understood in a highly abstract sense. For example, in his first paper on QED, Feynman considered his new solution for the propagation of the electron in external potentials, according to the Dirac equation:

> In this solution, the "negative energy states" appear in a form which may be pictured (as by Stueckelberg) in space time as waves traveling away from the external potential backwards in time. Experimentally, such a wave corresponds to a positron approaching the potential and annihilating the electron...Equivalence to the second quantization theory of holes is proved in an appendix.[16]

Since one can draw such a "picture" of backward-moving waves on a space-time diagram, or their corresponding ray representations, there seems to be a kind of visualization taking place, but as Dyson stressed, all that is important is the topological connections of the diagram, and in four-momentum space, which is where physicists generally use the diagrams as an algorithm, there *is* no time, but we are back once again to negative energies.

Nevertheless, there is no question that Feynman diagrams introduced a powerful new and simpler way of thinking about what would otherwise be highly abstruse and difficult matters, especially those involving virtual states. The Feynman diagram has been compared to the silicon chip in terms of its efficacy as a computational device.[17] Feynman himself always stressed the ensemble of *all* paths representing a given process, and he expanded the (Green's function) solutions of the equations of motion, rather than dealing with the equations themselves. The "trajectory" in the space-time diagram is representative of an uncountable infinity of such paths, considered to be followed simultaneously by the particle. According to Yoichiro Nambu, this concept "seems somewhat strange at first look and resists our minds that are accustomed to causal laws." Nambu continued:

> The...view of the entire space-time behavior of nature *sub specie aeternitatis*, however, might not appeal to a reason which is liable to think in the language of differential equations and pursue the development of things along a certain parameter. In fact we find it hard to regard the world line of a particle as a mere status of that particle, but are unconsciously following the motion of an imaginary point along a world line. Thus, in Feynman's theory where the ordinary time loses its role as an indicator of the development of the world, it would still be convenient to introduce some parameter with which the four-dimensional world is going to shape itself.[18]

That parameter is typically the proper time, long known in relativistic physics, but the phrase "backward-in time" was bound to lead to the kind of general intellectual excitement that the description of Einstein's kinematical and gravitational theories as "relativity" led to in earlier years.

In characterizing Feynman as a "magician," Kac suggested that a very powerful intuition was at work. There is little doubt that Feynman was very confident in his conceptualization, but on the other hand, he relied heavily on working out many examples to test each idea, a pursuit at which he was unparalleled in excellence, and also in industriousness. And while he feigned contempt for mathematical rigor, he paradoxically deserved to be called a "conscience" of physics. These points are illustrated by this remark in his Nobel address:

> In the face of the lack of direct mathematical demonstration, one must be careful and thorough to make sure of the point, and one should make a perpetual effort to demonstrate as much of the formula as possible. Nevertheless, a very great deal more truth can become known than can be proven.[19]

A second illustration of Feynman's conscientiousness is provided by the notion of "regularization." The presence of unacceptable divergences in QED led in the thirties to arbitrary prescriptions being advanced to eliminate them, the so-called "subtraction physics."[20] One of the major problems was that this involved the subtraction of one infinite quantity from another to get a finite remainder, a mathematically ambiguous operation. It is not often emphasized that although the first renormalization results of Julian Schwinger and Sin-itiro Tomonaga (who shared the Nobel Prize with Feynman) were improvements, in the sense that the requirement of relativistic covariance removed *some* of the arbitrariness of the subtraction procedure, they still amounted to the subtraction of infinite quantities. Whereas, it was the "intuitive" Feynman who introduced a consistent covariant procedure, namely a relativistic cutoff in the form of an additional large regulator "mass." Feynman's modified theory was finite, and after the necessary finite subtractions were made, the results remained finite in the limit of infinite regulator mass, whose effects then became unobservable. This procedure, analogous to the "summability" or "integrability" definitions employed in mathematical analysis, are mathematically unexceptionable. (Physically, however, this procedure was criticized by Dirac on the grounds that the theory with finite cutoff was not gauge invariant.) Subsequently, the regularization method was improved by Pauli and Villars, and by others.

Perhaps the deepest and most lasting contribution of Feynman to theoretical physics was his path-integral formulation of quantum dynamics. This is a genuine alternative to the Schrödinger and Heisenberg formulations of quantum theory. It is also distinct from Dirac's transformation theory of quantum dynamics, although it is based on a seminal observation of Dirac. The method was developed originally in Feynman's doctoral thesis, and Feynman described it as a means of avoiding the Hamiltonian formulation, which inevitably results in a picture of development in real time. As Feynman searched for a method incorporating an overall space-time view, he was led to consider a formulation in terms of the action principle. This is a generalization of the classical principle of least action, and it can be used to characterize the solutions of any quantum mechanical problem, although it is formulated as a principle connecting initial and final states. (It is not directly applicable to stationary states, but these can be treated as the solutions of the usual wave equations, which are derived from Feynman's action principle.) Dyson was the "prophet" of Feynman who related Feynman's method to that of the Heisenberg S matrix and labeled it as sum-over-histories.[21]

The path-integral method has proven invaluable in the theories of quantum mechanics of particles and fields, condensed matter, and statistical mechanics. It was especially valuable in proving the renormalizability of the theories of gauge fields which presently dominate the theory of elementary particles (the so-called Standard Model) and in treating applications of the modern "effective field theories." For example, in applying QCD to the wide range of problems of what was formerly known as high energy physics, but is now called medium-energy physics, such as pion scattering, nuclear forces, nuclear magnetic moments, etc., a path-integral algorithm known as *lattice gauge theory* is employed with the help of supercomputers.

Dick Feynman was fascinated with the new insights into cognition, reasoning, and other psychological issues raised by the theories of computer science. As the keynote speaker at a conference dealing with these and other problems, he discussed the possibilities of simulating physics, classical and quantum physics, by means of a universal (conceptual) computer—not a physical problem, but physics itself, i.e., the world as seen by physics. He began a section headed *Simulating Probabilities* as follows:

> Turning to quantum mechanics, we know immediately that here we get only the ability, apparently, to predict probabilities. Might I say immediately, so that you know where I really intend to go, that we have always had (secret, secret, close the doors!) a great deal of difficulty in understanding the world view that quantum mechanics represents. At least I do, because I'm an old enough man that I haven't got to the point that this stuff is obvious to me. Okay, I still get nervous with it. And therefore, some of the younger students...you know how it always is, every new idea, it takes a generation or two until it becomes obvious that there's no real problem. I cannot define the real problem, therefore I suspect there's no real problem, but I'm not sure there's no real problem. So that's why I like to investigate things.[22]

In a similar vein, when asked in a radio interview on the BBC, how he felt about the standard Copenhagen interpretation of quantum mechanics (which was rejected by so many of the founding fathers of quantum theory: Planck, Einstein, de Broglie, Schrödinger, etc.), Feynman confessed that it made him uncomfortable. But he went on to say that his discomfort did not mean that there was really any problem with it. Twenty years hence, he said, there might not be anyone alive who would feel that there were any conceptual difficulties with it. In stating this view he was, perhaps unconsciously, echoing Max Planck, who in his *Autobiographical Notes* said that older physicists can never be convinced of fundamentally new ideas, which triumph only as the older generation dies off.[23]

Feynman was aware that great new scientific insights are almost always born in the minds of young investigators, and before he had reached the age of fifty, he was beginning to fear the prospects of "elder statesmanship." In an interview several years before his death, he said, "I'm an old man now," and he acknowledged the "foolish" tendency to brand the newest ideas in physics as wrong. Then Feynman went on to say, "I am going to be very foolish" and he proceeded to express his strong reservations about "this superstring stuff." At the same time, one malady that sometimes afflicts older physicists was absolutely rejected by Feynman: the notion that the end of physics was in sight—that everything could be neatly expressed in a Standard Model. To the end of his life, Dick maintained that physics was an open quest. "Today," he said, "there are a large number of things that are not understood. That isn't fully appreciated, and people think they're very close to the answer, but I don't think so."

Physics was the core of Feynman's life. He loved to think and talk about it, and he never could understand why everyone did not find the same joy that he found in the wonders of the natural world. Feynman gave expression to this puzzlement in his famous *Lectures* in a revealing and eloquent way. In the third lecture, Feynman found himself saying, "...the stars are made of atoms of the same kind as those on the earth." But he could not leave it at that. A footnote followed: "How I'm rushing through this." Then he explained:

> Poets say science takes away from the beauty of the stars—mere globs of gas atoms. Nothing is "mere." I too can see the stars on a desert night, and feel them. But do I see less or more? The vastness of the heavens stretches my imagination—stuck on this carousel my little eye can catch one-million-year old light. A vast pattern—of which I am a part—perhaps my stuff was belched forth from some forgotten star, as one is belching there. Or see them with

the greater eye of Palomar, rushing all apart from some common starting point when they were perhaps all together. What is the pattern, or the meaning, or the *why*? It does no harm to the mystery to know a little about it. Far more marvelous is the truth than the artists of the past imagined! Why do poets of the present not speak of it? What men are poets who can speak of Jupiter if he were like a man, but if he is an immense spinning sphere of methane and ammonia must be silent?[24]

The people of the world are poorer, now that death has silenced Dick Feynman.

References

1. Mark Kac, *Enigmas of Chance: An Autobiography* (New York: Harper & Row, 1985), p. xxv: "An ordinary genius is a fellow that you and I would be just as good as, if we were only many times better. There is no mystery how his mind works...It is different with magicians...the working of their minds is for all intents and purposes incomprehensible. Even after we understand what they have done, the process by which they have done it is completely dark...Richard Feynman is a magician of the highest caliber."

2. *Surely You're Joking Mr. Feynman*, Richard Feynman with Ralph Leighton (New York: Norton, 1985).

3. Feynman put forth his claim to social irresponsibility in an article on Los Alamos in 1976; the same article was included in Ref. 2. Schweber's comments are in S. S. Schweber, "Feynman and the visualization of space-time processes," *Rev. Mod. Phys.* **58** (1986), pp. 449–508, especially p. 467. On the same page, Schweber quoted a letter written about Feynman by Robert Oppenheimer in 1944 to R. T. Birge, the chairman of the Physics Department at Berkeley: "He is not only an extremely brilliant theorist, but a man of the greatest robustness, responsibility, and warmth, a brilliant and lucid teacher, and an untiring worker...He is one of the most responsible men I have ever met. He does not regard himself as a privileged artist but as one of a group of hard working men for whom the development of physical science is an obligation, and the exposition both an obligation and a pleasure."

4. Feynman's own most complete account of that investigation is a chapter in his book *What Do You Care What Other People Think* (New York: Norton, 1988).

5. *The Feynman Lectures on Physics*, Richard P. Feynman, Robert B. Leighton, and Matthew Sands (Reading, Mass.: Addison-Wesley, 1963), Vols. I–III.

6. Ref. 5, Vol. I, p. 1–1.

7. Ref. 5, Vol. I, p. 1–2.

8. Ref. 5, Vol. I, p. 1–8.

9. Ref. 5, Vol. I, p. 1–9. In a similar vein, he stated on p. 3–6: "Poets say science takes away from the beauty of the stars—mere globs of gas atoms. Nothing is 'mere.' "

10. Ref. 5, Vol. I, p. 2–1.

11. Ref. 5, Vol. I. p. 3–9.

12. "The development of the space-time view of quantum electrodynamics," *Physics Today* **19**, August 1936, pp. 31–44.

13. Ref. 12, p. 32.

14. J. A. Wheeler and R. P. Feynman, "Interaction with the absorber as the mechanism of radiation," *Rev. Mod. Phys.* **17** (1945), pp. 157–181, and "Classical electrodynamics in terms of direct interparticle action," *ibid.* *21*, pp. 425–433. These are reprinted in an interesting collection, *The Theory of Action-at-a-Distance in Relativistic Particle Dynamics*, ed. Edward H. Kerner (New York: Gordon and Breach, 1972). The editor's *Preamble* says of theories of delayed action-at-a distance: "A major example...is Wheeler and Feynman's electrodynamics, in which a point on one particle's world line reaches out forward and backward along the double light-cone to touch with equal weight upon both advanced and retarded world points on another's world line. The electro-magnetic touch is directly through the Lienard-Wiechert *solutions* to Maxwell's equations, so the world-line dynamics involves only world-line variables...the time-symmetrical interaction (with explicit omission of self-interaction) allows an action principle (Fokker) and conservation laws; and the further hypothesis...of 'complete absorption' in the electromagnetic universe is a kind of electrodynamic Mach's principle accounting marvelously for the appearance of the Lorentz-Dirac force of radiation damping, and for the appearance of retarded interactions, on the local scene."

15. Ref. 14, "Classical electrodynamics in terms...", p. 27.

16. R. P. Feynman, "The theory of positrons," *Phys. Rev.* **76** (1949), pp. 749-759, especially p. 749.

17. Schwinger made this comparison at a Fermilab symposium in May 1980. In describing the course of his own thinking in 1949, after Feynman's methods had begun to be widely used, he said: "It is only human that my first action was one of reaction. Like the silicon chip of recent years, the Feynman diagram was bringing computation to the masses. Yes, one can analyze experience into individual pieces of topology. But eventually one has to put it all together again. And then the piecemeal approach loses some of its attraction." J. Schwinger, "Renormalization theory of quan-

tum electrodynamics: an individual view," in *The Birth of Particle Physics*, ed. L. M. Brown and L. Hoddeson (Cambridge University Press, 1983), especially p. 343.

18. Y. Nambu, "The use of the proper time in quantum electrodynamics," *Prog. Theor. Phys.* **5** (1950), p. 82.

19. Ref. 12, p. 43.

20. Many sarcastic references to this kind of physics in the 1930s are to be found in *Wolfgang Pauli, Scientific Correspondence, Vol. II.*, ed. by K. von Meyenn (Berlin: Springer-Verlag, 1985). E.g., in Pauli to Heisenberg, 1 Nov. 1934, on p. 357: "The present situation in theoretical physics is this, that a subtraction physics of electrons and positrons stands beside a nuclear physics of undetermined functions."

21. A clear pedagogical explanation of this method is in R. P. Feynman and A. R. Hibbs, *Quantum Mechanics and Path Integrals* (New York: McGraw-Hill, 1965).

22. R. P. Feynman, "Simulating physics with computers," *Intern. Journ. of Theoretical Physics* **21**, (1982), pp. 467–488, especially p. 471.

23. Planck's words were: "A new scientific truth does not triumph by convincing its opponents and making them see the light, but rather because its opponents eventually die, and a new generation grows up that is familiar with it." Max Planck, *Scientific Autobiography*, trans. by F. Gaynor (New York: Philosophical Library, 1949), p. 33.

24. Ref. 5, Vol. I, p. 3–6.

The Early Years

THE YOUNG FEYNMAN

John Archibald Wheeler

"'This chap from MIT: Look at his aptitude test ratings in mathematics and physics. Fantastic! Nobody else who's applying here at Princeton comes anywhere near so close to the absolute peak." Someone else on the Graduate Admissions Committee broke in, "He must be a diamond in the rough. We've never let in anyone with scores so low in history and English. But look at the practical experience he's had in chemistry and in working with friction."

These are not the exact words, but they convey the flavor of the committee discussion in the spring of 1939 that brought us 21-year-old Richard Phillips Feynman as a graduate student. How he ever came to be assigned to this 28-year-old assistant professor as grader in an undergraduate junior course in mechanics I will never know, but I am eternally grateful for the fortune that brought us together on more than one fascinating enterprise. As he brought those student papers back—with errors noted and helpful comments offered—there was often occasion to mention the work I was doing and the puzzlements I encountered. Discussions turned into laughter, laughter into jokes and jokes into more to-and-fro and more ideas.

John Archibald Wheeler is Joseph Henry Professor of Physics, Emeritus, Princeton University, and Ashbel Smith and Jane and Roland Blumberg Professor, Emeritus, University of Texas at Austin.

The Busted Bottle

One day our discussions led to Mach's principle. We knew of the inspiration Einstein had found in thinking of inertia as originating in acceleration—not relative to Newton's absolute space, but relative to Mach's faraway stars. Was it a problem in the junior course in mechanics that started us thinking about the familiar lawn sprinkler? Shaped like a swastika, it shoots out four jets of water. The recoil drives the sprinkler arms round and round. But where does the recoil act? Doesn't it act at the point where the stream of water suddenly changes direction from straight out to straight transverse? But suppose the arm sucks water in instead of squirting it out. Surely, we said to each other, there is an identical change in direction and therefore an identical reaction. Surely the sprinkler will again turn round when water in the arms is sucked in rather than being shot out. Oh no, it won't. Oh yes, it will. We had a great time trying out both sides of this question on our colleagues. As the days went by, more and more colleagues up and down the corridors took positions. The debate grew more animated. No argument of theory was strong enough to still the disagreements. The situation called for an experiment.

Feynman made a six-inch miniature lawn sprinkler out of glass tubing and hung it from a flexible tube of rubber. He checked that it worked OK as a sprinkler. Then he wangled the whole dangling gadget through the throat of a great glass carboy filled with water. He got this outfit set up on the floor of the cyclotron lab, where there was a handy compressed-air outlet. He ran the compressed air in through a second hole in the cork at the top of the carboy. Ha! A little tremor as the pressure was first applied, as water first began to run backward through the miniature lawn sprinkler. But, as the flow continued there was no reaction. Then increase the air pressure. Get more backward flow of water. Again a momentary tremor at the start of this maneuver but no continuing torque. OK, more pressure. And more! Boom! The glass container exploded. Water and fragments of glass went all over the cyclotron room. From that time onward Feynman was banished from the lab.

Everything as Scattering

I enlisted Feynman's help on one of the ever-expanding problems I had brought back to Chapel Hill and then to Princeton from my post-doc days. At the great Rutherford-centered October 1934 London-

Cambridge International Conference on Physics, four puzzles stood out. Of them none excited me more then and in my subsequent Copenhagen year than the problem of the mini-shower, as I called it—the puzzle of the so-called "anomalous" back scattering of gamma rays by lead. Almost every elementary process of photon physics was needed to understand the 1930–35 experimental results of Louis Gray and Gerald Tarrant, of Chung-Yao Chao, Lise Meitner and H. H. Hupfeld and of Jacob Jacobsen: production of Compton electrons, photoelectrons and pair electrons by the incident 2.6-MeV gamma ray, and electron—and photon—scattering, both single and multiple. For each elementary process I had a symbolic diagram and a curve of cross section as a function of energy; but to combine these processes into a prediction about the spectrum of back-scattered radiation much numerical slogging was needed.

As Feynman and I reviewed that enterprise we found we didn't have the heart for it. It remains undone to this day. Instead, we found ourselves entranced by two issues peripheral to the original undertaking: What is the complete story of Compton scattering within the framework of the Fermi-Thomas statistical-atom model? And how could we understand, in terms of scattering and nothing but scattering, the propagation of a photon through a medium of variable refractive index, or the passage of an electron through a position-dependent atomic potential? How many wonderful aspects of physics came together in these two enterprises, especially the second one: Huygen's principle as concept of how light (and—in our day—matter) propagate; refractive index as the cumulative consequence of many individual scattering processes; spirals—Cornu and other—as tool to add up scattered waves; and as a motto to inspire us, the phrase "everything as scattering." What fun it was, what jokes along the way, what a happy mix of diagrams and equations, of the well known and the new! That work never got published but both of us went on in postwar years to capitalize on the insights we had won from it.

The Tumbling Can

Sometimes we worked together in my Fine Hall office, three blocks east of Feynman's Graduate College room where he lived and worked—writing away hour after hour on one of those fan-folded pads of computer print-out paper, as enormous then as now. But for a long stretch of pow-wow, lasting two or three hours, we generally worked at my house two blocks west of the Graduate College. As we came downstairs from the work room for supper, Letitia, five, and Jamie, three, would follow him, hoping for one of the jokes or tricks he usually had up his sleeve. As those bright eyes tagged along, he teased, "A tin can." He came into the kitchen where my wife was cooking dinner and took off the counter a can not yet opened. "A tin can: I can tell you whether what's inside is solid or liquid without even opening it or looking at the label. Do you know how?"

"How?" came the response from the little people.

"By the way it turns when I toss it up in the air." And toss it he did, in an arc of wild precession. "Liquid," he announced. We could all see his prediction checked out right when the can was opened.

Hypnotized?

"Be my guest at the next Wednesday night dinner at the Graduate College," Feynman suggested one day. "There's going to be a talk on hypnotism and a demonstration." When the call came for a volunteer, it was Feynman who stood up and went up to the front of the crowded room. The hypnotist made his motions, spoke his abracadabra. In a sepulchral voice he intoned his instructions: "Walk to the corner of the room. Turn. Pick up the book that you will find lying before you. Balance it on your head. Bring it to me." Feynman, looking like a sleepwalker, performed as commanded. He went on to fulfill further instructions. At last he was released.

Knowing Feynman, and having watched his performance, I came to an everyday, matter-of-fact theory of "hypnotism": It's acting. The Shakespearean player is animated to act his demanding part by the subtle pressure of the expectations of those around. So in hypnotism! Given an unfamiliar part to act, no one I know ever rose to the challenge more delightedly, more imaginatively and with more fun to his audience than Richard Feynman.

It would be tempting, if space permitted, to go on from the case of the tumbling can and the hypnotic trance to other stories of life with Feynman at Princeton: the black box electric circuit, the quaking jellyfish and the anodized-iron memory device. The last two are precursors, surely, of his lifelong interest in the mechanism of brain action. That interest showed never more clearly than in the Caltech seminar Feynman taught in his last years, first jointly with John J. Hopfield and

KAISER WILHELM-INSTITUT FÜR CHEMIE
Professor Dr. LISE MEITNER

FERNSPR.: G 4. BREITENBACH 2391 u. 2362   BERLIN-DAHLEM, DEN 23.März 1935.
                                                      THIEL-ALLEE 63

Herrn I.A. W H E E L E R,

Institut für Theoretische Physik,

K O P E N H A G E N.

Lieber Herr Wheeler,

Leider ist Dr.v.Droste, der die Streuungsmessungen an
der γ-Strahlung bei 60° gemacht hat, derzeit krank, so dass
ich Ihnen jetzt nichts über die Einzelheiten der Kurven berich-
ten kann. Ich schicke Ihnen gleichzeitig die Arbeit von   —
Dr.Kösters und hoffe, Ihnen nächste Woche, wenn Dr.v.Droste
wieder im Institut ist, auch etwas näheres über dessen Messungen
schreiben zu können.

Mit besten Grüssen

Lise Meitner

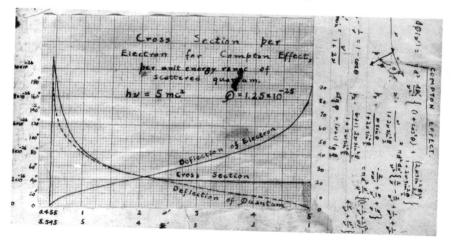

Letter and notebook page from 1935, subject of 1939–41 discussions with Richard Feynman. The letter (above), from Lise Meitner to the author, refers to the latest results on the "anomalous" backscattering of 2.6-MeV gamma rays. The page (below), torn out of the author's workbook of the time, refers to the Compton effect, and is a sample of the data sheets of the many elementary processes that come into play in his minishower interpretation of the phenomenon. Standard diagrams of this type—the subject matter of consultations between Feynman and Wheeler—have a little of the flavor of ideological antecedents of the later, far more abstract, Feynman diagrams.

Carver Mead, and then on his own (see the articles by David L. Goodstein on page 115 and by W. Daniel Hillis on page 139).

Precursor to a Thesis Topic

Richard Feynman is one of the many wonderful thesis advisees who, over the years, have done so much to teach me. In expressing indebtedness to him for many an insight, I testify also to the immense gratitude I feel to all students who have instructed me.

In 1939 Feynman had not yet decided what he was going to work on for his thesis or with whom. As a graduate student not yet committed to any particular topic or adviser, and being free—like all Princeton graduate students in physics then and now—of all formal course requirements, he had spread out before him all the richness in mathematics and physics of the university and the Institute for Advanced Study. He knew that I, on the other hand, was a divided man, torn between all the commitments that came in the wake of my fission work and an unquenchable curiosity about the foundation problems of physics. From more than one of my courses he knew my faith that whatever is important is at bottom utterly simple. But wasn't my 1934–35 idea crazy: to out-Dirac Dirac and count the electron as the basis of everything, of all particles, of the so-called "strong forces" of the nucleus, of even the electromagnetic field? Yet Feynman expressed some interest in this idea—and more in having even busy me as his adviser.

Interaction with the Absorber

Animated by the concept of "everything as electrons," I took time off from more immediate concerns one Sunday afternoon in the sunlit upstairs work room at home, and figuring on the back of an envelope I discovered that I could give a quantitative account of radiative reaction in terms of forces produced by the particles of the faraway absorber. The density of those particles and their distance—it turned out—cancel out provided only that there are enough particles around to guarantee complete absorption of the outgoing radiation. However, the strength I got this way for the force of radiative reaction was off by a factor of two from the well-known and often tested value.

The next morning, when Feynman came in with the homework to be returned to the students, I told him about my finding and my difficulty with the factor two. He jumped into the middle of this new game with his usual vim. He soon spotted the source of the trouble—I had undercounted the effective force exerted by the emitter on the absorber. Then all fell into place.

Not long afterward we gave a seminar report on our finding. At tea time a few days later Wolfgang Pauli said to me in a worried way that he felt that our result arose somehow from some mathematical tautology. However, Feynman and I went around to Einstein's house at 112 Mercer Street to talk to him about our work. We found him both interested and sympathetic. He told us about a paper he had written with Walter Ritz to record their disagreement on the mechanism of radiation damping—to us a wonderful example of true colleagueship and of responsibility in the realm of science. In this brief paper Ritz argued that the irreversibility of radiative reaction is a consequence of some irreversibility in electrodynamics itself. Einstein took the opposite position. In his view all truly basic equations for the dynamics of particles and fields are in and by themselves invariant with respect to reversal of the direction of time. The damping, in Einstein's view, originated somehow from asymmetry in the initial conditions. He expressed a strong interest in our work because we had at last given a concrete picture of what those initial conditions are and how they work.

Not until after the war, in stolen hours at conferences in Los Alamos and elsewhere, did we have the opportunity to present[1] an outlook so novel with some of the care it required.

A New Method for a Problem of a New Kind

Our concept of direct action at a distance between charged particles, without the intermediation of any field: how to translate it from classical theory to quantum theory? How to capitalize for this purpose on the action principle of Adriaan Fokker? Feynman, with his wonderful zip, grabbed this issue and ran with it. A hint, in a paper of Paul Dirac, Feynman had soon magnified up into a complete prescription for quantization,[2] his famous method of "sum over histories" or "path integration"—also written up in full and published[3] only after the war.

Phase as it came into the scattering problems that we had been considering, phase as it comes into the time-dependent Schrödinger wavefunction, phase as seen in Feynman's wonderful new method of sum over histories! To see this central place of wave phase in the scheme of things was to see in a new light the central place of the action principle in classical mechanics. I was learning from these discussions with Feynman that the integrated action of classical theory, in a sense more precise than ever before appreciated, is—apart from a universal factor, $\hbar = 1.054 \times 10^{-27}$ g cm^2/sec—only another name for the phase of the probability amplitude associated with the classical history.

Visiting Einstein one day, I could not resist telling him about Feynman's new way to express quantum theory. "Feynman has found a beautiful picture to understand the probability amplitude for a dynamical system to go from one specified configuration at one time to another specified configuration at a later time. He treats on a footing of absolute equality every conceivable history that leads from the initial state to the final one, no matter how crazy the motion in between. The contributions of these histories differ not at all in amplitude, only in phase. And the phase is nothing but the classical action integral, apart from the Dirac factor, $\hbar$. This prescription reproduces all of standard quantum theory. How could one ever want a simpler way to see what quantum theory is all about! Doesn't this marvelous discovery make you willing to accept quantum theory, Professor Einstein?" He replied in a serious voice, "I still cannot believe that God plays dice. But maybe," he smiled, "I have earned the right to make my mistakes."

Undeterred I persisted, and still do, in regarding Feynman's PhD thesis as marking a moment when quantum theory for the first time became simpler than classical theory. I began my upcoming graduate course in classical mechanics with Feynman's idea that the microscopic point particle makes its way from A to B, not by a unique history, but by pursuing every conceivable history with democratically equal probability amplitude. Only out of Huygens's principle, only out of the concept of constructive and destructive interference between these contributions—and this only in an approximation—could one understand the existence of the classical history. Feynman sat there and took the course notes, of which I still have a mimeographed copy. On many a puzzling point he helped us both to find new light by discussions in class and out.

Any Career for the Kid from Far Rockaway?

While Richard was working on his thesis, his father, Melville Arthur Feynman, sales manager for a medium-sized uniform company, made a brief call on me in my office one day. How important he had been in

Feynman's upbringing many of us saw in a Feynman television program,[4] and more of us can read in his two autobiographical bestsellers.[5] The father was concerned whether his son had any future. "A brilliant one," I assured him. "But won't he be handicapped by his simple background, or maybe even by some kind of anti-Jewish prejudice?" "No," I replied, and went on to describe the career histories of several close colleagues. I did not tell him that in college days in Baltimore I had been one of the founders and first president of the lively Federation of Church and Synagogue Youth!

From Student-Teacher to Customer-Supplier

Concern about the imminence of war drew some of our Princeton colleagues to the MIT Radiation Laboratory. Simultaneously the uranium work at Princeton grew: Heinz H. Barschall, Morton Kanner and Rudolf Ladenburg were doing controlled-neutron-energy experiments; Edward Creutz, Lewis A. Delsasso and Robert Wilson were working with the cyclotron; and Henry De W. Smyth, Louis A. Turner, Eugene Wigner and I were doing theoretical analysis. We brought Feynman into this work. Some months after Pearl Harbor some of us, including him, moved to Los Alamos, where Turner's plutonium concept was destined to win an ever bigger place. Before going Feynman took his final PhD oral. I was sad to have to miss it. However, I had already been called to Chicago to forward the uranium project. By fall, the West Stands pile—the first nuclear chain reactor—was on its way to final assembly, and Arthur Compton asked me to take Chicago know-how to Du Pont, manufacturer of the plutonium for our customer, Richard Feynman's Los Alamos. More than once he and I had to meet at Los Alamos to help formulate meticulously reliable safety precautions for the chemical separation of the plutonium at the Hanford plant.

One night Richard and I went out on the mesa with Joe Fowler and team to witness a high explosive implosion test. And at the lab with what enthusiasm he explained to me how he had found out that heat can't be hidden. Over and over on one of the best of the card-controlled IBM computers of that time he had calculated the same hydrodynamic implosion run. The resulting motions and pressures here and there in the metal came out filled with frightful irregularities. These irregularities, moreover, came out totally different from one run to the next. What was wrong with the computer? Suddenly he had grasped its message. The program had failed to include a term for heat. The machine knew better. If the stupid equations weren't going to include heat, the computer would have to impose its own way to represent heat: motion varying chaotically from point to point and from instant to instant. With what zest he explained this, and what delight he had in the nighttime fireworks! But all the time I knew the burden that lay close to his heart.

His wife, Arline, lay slowly dying in the hospital at Albuquerque.

Arline

On one of these trips from Hanford, Washington, to Los Alamos, New Mexico, I took the occasion to visit Arline in the hospital. My wife and I had first come to know Arline Greenbaum when Richard invited her down from his town, Far Rockaway, New York, to Princeton for one or another of the occasional Saturday night dances at the Graduate College. Auburn-haired, Arline was not attractive; she was *very* attractive. Two lively watercolors by her we gained as souvenirs of these special weekends.

Those Princeton dances were respites from her overtaxed life. She was a full-time art student in New York by day. By night she taught piano, earning the wherewithal to pay for those art lessons. The strain of the double life in time, I believe, proved too much. She picked up an infection. Months went by with divided doctors diddling with their diagnoses. When finally she was coughing blood and it was unmistakably tuberculosis, it was too late. Feynman's latest book, which he did not live to see, tells the affecting story of how the two young people, against the advice of family and friends, and knowing early death was certain, nevertheless married, shared all their deepest concerns and stood devotedly together until the end. A week after I said good-bye to her hospital, her oxygen line and Arline herself, she was gone.

Arline was a strong character. She was one of the few people I have known who could stand up to Richard. She and his father and mother were guides he trusted. It was Arline who gave him the advice that forms the title of his last book, *"What Do You Care What Other People Think?"*

References

1. J. A. Wheeler, R. P. Feynman, Rev. Mod. Phys. **17**, 157 (1945); Rev. Mod. Phys. **21**, 425 (1949).

2. R. P. Feynman, "A Principle of Least Action in Quantum Mechanics," PhD thesis, Princeton Univ. (1942).

3. R. P. Feynman, Rev. Mod. Phys. **20**, 367 (1948). R. P. Feynman, A. R. Hibbs, *Quantum Mechanics and Path Integrals*, McGraw-Hill, New York (1975).

4. Nova, "The Pleasure of Finding Things Out," broadcast 25 January 1983.

5. R. P. Feynman, as told to R. Leighton, *"Surely You're Joking, Mr. Feynman!"* Norton, New York (1985). R. P. Feynman, as told to R. Leighton, *"What Do You Care What Other People Think?"* Norton, New York (1988).

This article was originally published in the February 1989 issue of *Physics Today*.

At Los Alamos

FEYNMAN IN LOS ALAMOS AND CORNELL

Hans Albrecht Bethe

===

Richard Feynman came into my life at the beginning of the Los Alamos Laboratory, in April 1943. In the first few days, we already competed in calculational tricks. I showed him how to quickly calculate the square of numbers $50 \pm x$, where x is between 1 and 9. He answered with a much more important invention, namely how to integrate numerically a third order differential equation. That method is very accurate, and was used in a theory of Robert Serber on the calculation of the yield of nuclear weapons.

Serber's theory applied mainly to weapons slightly above critical mass. One evening, Feynman and I got together to invent a formula for larger masses. This Bethe-Feynman formula was quite accurate, as it later turned out, and it continued in use as a quick approximation for decades. It was a pleasure to work with him.

I soon made Feynman a group leader; he was the youngest of them at age 25, most of the others were about ten years older. His group was charged with the calculation of the behavior of uranium hydride in a nuclear explosion. This had been suggested by Edward Teller because

Hans Albrecht Bethe is John Wendell Anderson Professor Emeritus of Physics at Cornell University.

the hydride has a much lower critical mass than uranium metal. However, the calculations gave an unsatisfactory result: Because the neutrons are very much slowed down by the hydrogen, the rate of neutron multiplication is low, and therefore the yield of such an explosion is also low. However, the work of the group led to useful methods of calculation which were later used by at least one member of the group in calculating nuclear reactors with a water moderator.

In addition to these duties, Feynman spent a lot of time repairing typewriters and desk calculators. He loved mechanical things, and there was no typewriter repairman at all in the laboratory. However, I thought this was not a good use of his ability and told him to stop.

Soon a greater mechanical challenge came. The laboratory acquired an IBM computer which came in many boxes, disassembled. Feynman and Eldred Nelson immediately got to work using the instructions that came with the machines. In two or three days they had all eight or ten machines assembled. A week later, there came an official IBM repairman who had been drafted by the Army and assigned to our laboratory. He said that he had never seen the machines assembled by nonexperts, and apparently all had been assembled correctly.

A group was set up to carry out the calculations on the IBM. The two leaders of the group were fascinated with the capabilities of the machines and loved to play with them. But the important calculations for our project did not get done. After several months of this, I decided that something had to be done. The only man in my division who was obviously capable of doing this was Richard Feynman so I made him the leader of the IBM group. Soon the group was working efficiently, and we got our answers promptly and steadily.

In his spare time, Feynman used to engage in opening the safes of various members of our division. This caused great amusement and was sometimes useful. He has discussed this himself in his posthumous book, "What Do You Care What Other People Think?," edited by Ralph Leighton.

Richard's wife Arline had incurable tuberculosis and was in a hospital in Albuquerque. (This was described in the Leighton book.) He had a deep love for Arline. He visited her every weekend, which was difficult because he did not have a car. Somehow he managed to catch a ride for the nearly 100 miles from Los Alamos to Albuquerque and back. In between, he invented codes and wrote letters to Arline in code so that she would have some mental work to entertain her. These letters annoyed the censor of our mail, even after Feynman enclosed the code with each letter.

I thought it would be very good to have Feynman at Cornell after the War. I wrote a letter to our department chairman, and the appointment was soon accomplished. Feynman went to Cornell very soon after the end of the War, and was there for the beginning of the Fall semester of 1945. I returned in time for the Spring semester of 1946. For some period we worked jointly on problems of large air showers of cosmic rays, but Feynman was somewhat at loose ends to find a worthwhile subject. I accepted a consulting agreement with the General Electric Research Laboratory to work on nuclear physics; later this became the Knolls Atomic Power Laboratory, concerned with the design of nuclear reactors, but in the summer of 1947 we could still do pure nuclear physics. Feynman also became a consultant. During this time, he was called for a physical examination with the purpose of drafting him into the Army. In a hilarious session, the psychiatrist of the examination board declared him unfit. When he came back to General Electric and told me the details, he and I laughed uproariously for half an hour.

A decisive event for Feynman's work was the Shelter Island Conference in June 1947. About 30 theoretical physicists attended that conference, and we exchanged information on the latest discoveries. Two matters in particular excited all of us: One was the idea of renormalization which had been pursued, particularly by Hans Kramers, and which seemed able to remove the infinities from quantum electrodynamics. The other consisted of two experiments: that of Willis Lamb and Robert Retherford on the shift of the $2s$ level of atomic hydrogen, and the measurement by Polykarp Kusch of an apparently increased magnetic moment of the electron.

Many theoretical physicists set to work to explain these two experiments. Feynman went back to methods he had invented previously, which involved the calculation of the wave function at a time t if it was known at an earlier time 0, i.e., essentially using a Green's function. This was the basis of Feynman diagrams, now used generally by physicists to describe the behavior of particles. Feynman's Green's function was first developed for the nonrelativistic Schroedinger equation; the Dirac equation presented some difficulties. But Feynman solved these, inventing the idea of positrons being electrons moving backward in time. In a few months he had a complete relativistically invariant theory which could be renormalized.

In Spring 1948, Oppenheimer organized a repetition of the Shelter Island Conference, meeting at Pocono Manor. The centerpiece of this conference was the new development of quantum mechanics with renormalization. Julian Schwinger gave a beautiful lecture on his formulation. It lasted many hours. He led us logically from the usual quantum mechanics to his theory which was manifestly relativistically covariant and which permitted renormalization of charge and mass. Everybody was impressed, including Niels Bohr who had come to this conference, but who had not been at Shelter Island.

Then Feynman talked, and presented his graphs going from one interaction to the next, and lines going forward and backward in time. Niels Bohr protested violently; he thought that Feynman had disregarded the Uncertainty Principle, using interactions at a given position, followed by propagation with given momentum. Lines going backward in time were even harder to accept. This was a great disappointment to Feynman, especially because at Los Alamos he was the one who had given Niels Bohr and his son very extensive briefings on the scientific progress of the laboratory.

After Pocono, Richard seemed rather unhappy, and I consoled him, assuring him that his theory was correct. (Later on, his main recollection of Pocono was that he had had very happy and fruitful discussions with Schwinger.) Both Schwinger and Feynman then calculated the Lamb shift. Both of them made the same mistake and got the wrong answer. So French and Weisskopf were actually the first to obtain the correct answer for the Lamb shift, using old-fashioned quantum mechanics.

A couple of years later, Michel Baranger, a graduate student at Cornell, returned to the problem, trying to get a more accurate join of the nonrelativistic with the relativistic part of the calculation. I don't remember whether Baranger was Feynman's student or mine, but we all three worked on this problem. We got a correction of about 10 MHz to the result.

I did not follow Feynman's work in detail as he was doing it, but Freeman Dyson did and described it in his book, "Disturbing the Universe." Dyson almost immediately showed the equivalence of the methods of Feynman and Schwinger, and extended the theory to show that renormalization gives finite results in quantum electrodynamics to any order of perturbation theory. When the theory was complete, I learned the techniques myself and checked that positrons moving backward really agreed in every detail with the clumsy earlier formulation. I spent the Fall term of 1948 at Columbia University and lectured on the Feynman method there, as well as giving a seminar on it at the Institute for Advanced Study in Princeton.

The Cornell Years

FEYNMAN AT CORNELL

Freeman J. Dyson

I lived in the same town as Richard Feynman for only one year, the academic year 1947–48, when he was a professor at Cornell and I was a graduate student. After that year I saw him only occasionally, mostly among crowds at scientific meetings. My most vivid memories of Feynman come from that first year, now 40 years ago, when I was 24 and he was 30. I had the good luck to know him when he was at the height of his creative powers, furiously struggling to complete the theory he called the "space-time approach," which afterwards became the standard approach for everybody doing calculations in particle physics.

Being a dutiful son, I sent my parents in England a weekly letter recording my adventures in America. My parents preserved the letters and I now have them in my hands. Instead of writing fake recollections of Feynman for this issue, I decided to use direct quotes from the letters in which he appears. These extracts are fragmentary and do not tell us much about Feynman's inner thoughts. Their virtue is that they are authentic, written within a few days of the events they describe, without editing and without hindsight.

I arrived in America in September 1947 to do graduate work in phys-

Freeman J. Dyson is a Professor of Physics at the Institute for Advanced Study in Princeton, New Jersey.

ics under the direction of Hans Bethe. So far as I can tell from the records, I had never heard of Feynman before I came to Cornell. It took me quite a while to discover what a great man he was.

I begin with the letter in which Feynman makes his first entrance, written two months after my arrival in America.

Cornell University
19 November 1947

Just a brief letter before we go off to Rochester. We have every Wednesday a seminar at which somebody talks about some item of research, and from time to time this is made a joint seminar with Rochester University. To-day is the first time this term that we are going over there for it. It is a magnificent day, and it should be a lovely trip; Rochester is due north of here, and we go through some wild country. I am being taken in Feynman's car, which will be great fun if we survive. Feynman is a man for whom I am developing a considerable admiration; he is the brightest of the young theoreticians here, and is the first example I have met of that rare species, the native American scientist. He has developed a private version of the quantum theory, which is generally agreed to be a good piece of work and may be more helpful than the orthodox version for some problems; in general he is always sizzling with new ideas, most of which are more spectacular than helpful, and hardly any of which get very far before some newer inspiration eclipses them. His most valuable contribution to physics is as a sustainer of morale; when he bursts into the room with his latest brainwave and proceeds to expound it with the most lavish sound effects and waving about of the arms, life at least is not dull. [Victor] Weisskopf, the chief theoretician at Rochester, is also an interesting and able man, but of the normal European type; he comes from Munich, where he was a friend of Bethe from student days.

The event of the last week has been a visit from [Rudolf] Peierls, who has been over here on government business and stayed two nights with the Bethes before flying home. He gave a formal lecture on Monday about his own work, which was very well received, and has been spending the rest of the time in long discussions with Bethe and the rest of us, at which I learnt a great deal. On Monday night the Bethes gave a party in his honour, to which most of the young theoreticians were invited. When we arrived we were introduced to Henry Bethe, who is now 5 years old, but he was not at all impressed, in fact the only thing he would say was "I want Dick. You told me Dick was coming," and finally he had to be sent off to bed, since Dick (alias Feynman) did not mate-

rialize. About half an hour later, Feynman burst into the room, just had time to say "So sorry I'm late. Had a brilliant idea just as I was coming over," and then dashed upstairs to console Henry. Conversation then ceased while the company listened to the joyful sounds above, sometimes taking the form of a duet and sometimes of a one-man percussion band.

Peierls was shown around the new and lavishly built Nuclear Laboratory, which will be opened formally with a party on Saturday. Feynman remarked that it was a pity to think that after all the work that the builders had put into building it, so little would probably be done by the people who lived in it; and Bethe said that only because of the steel shortage is any nuclear physics worth mentioning done in the United States. This may well be true; at any rate the most outstanding experiments in the world are at present being done at Bristol by [Cecil] Powell with no apparatus more elaborate than a microscope and a photographic plate.

Next a brief item one week later.

27 November 1947

The trip to Rochester last week was a great success. I went there and back in Dick Feynman's car with Philip Morrison and [Edwin] Lennox, and we talked about everything from cosmic rays downward.

On Saturday we had our great inaugural party for the Synchrotron Building. It was a great success; I played my first game of poker and found I was rather good at it; I won 35 cents. The synchrotron itself does not arrive for some time yet, so the building is still empty. The party consisted chiefly of dancing and eating; Bethe and Trudy Eyges danced together for about an hour, very beautifully, while Rose Bethe and Leonard Eyges exchanged disapproving glances.

Leonard Eyges was another graduate student working with Bethe. Trudy was his wife.

Now we jump three months.

8 March 1948

Yesterday I went for a long walk in the spring sunshine with Trudy Eyges and Richard Feynman. Feynman is the young American professor, half genius and half buffoon, who keeps all physicists and their children amused with his effervescent vitality. He has, however, as I have recently learned, a great deal more to him than that, and you may be interested in his story. The part of it with which I am concerned began when he arrived at Los Alamos; there he found and fell in love with a brilliant and beautiful girl, who was tubercular and had been exiled to New Mexico in the hope of stopping the disease. When Feynman arrived, things had got so bad that the doctors gave her only a year to live, but he determined to marry her and marry her he did; and for a year and a half, while working at full pressure on the Project, he nursed her and made her days cheerful. She died just before the end of the war. [Actually, as I later learned, Feynman and his first wife had met long before his Los Alamos days.]

As Feynman says, anyone who has been happily married once cannot long remain single, and so yesterday we were discussing his new problem, this time again a girl in New Mexico with whom he is desperately in love. This time the problem is not tuberculosis, but the girl is a Catholic. You can imagine all the troubles this raises, and if there is one thing Feynman could not do to save his soul it is to become a Catholic himself. So we talked and talked, and sent the sun down the sky, and went on talking in the darkness. I am afraid that at the end of it poor Feynman was no nearer to the solution of his problems, but at least it must have done him good to get them off his chest. I think myself that he will marry the girl, and that it will be a success, but far be it from me to give advice to anybody on such a subject.

Then, a week later, the shape of things to come begins to emerge.

15 March 1948

My own work has taken a fresh turn as a result of the visit of Weisskopf last week. He brought with him an account, rather garbled in transit, of the new Schwinger quantum theory which [Julian] Schwinger had not finished when he spoke at New York. This new theory is a magnificent piece of work, difficult to digest but with some highly original and undoubtedly correct ideas; so at the moment I am working through it and trying to understand it thoroughly. After this I shall be in a very good position, able to attack various important special problems in physics with a correct theory while most other people are still groping. One other very interesting thing has happened recently; our Richard Feynman, who always works on his own and has his own private version of quantum theory, has been attacking the same problem as Schwinger from a different direction, and has now come out with a roughly equivalent theory, reaching many of the same ideas independently; this makes it pretty clear that the theory is right. Feynman is a man whose *ideas are as difficult to make contact with as Bethe's are easy; for this reason I have so far learnt much more from Bethe, but I think if I stayed here much longer I should begin to find that it was Feynman with whom I was working more.*

On the Road to Albuquerque

But three more months go by without any mention of Feynman. He reappears only in the last letter from Cornell.

11 June 1948

Although our society is breaking up and many old friends have already departed, these last days are not at all lonely; the experimentalists are still hard at work on their synchrotron, and in the last week we have had one supper-picnic, one swimming expedition and one sailing expedition, all of which were very enjoyable. However, this week I start out for the West, and no doubt that will be great fun too. Incidentally, the American "picnic" is not exactly what we understand by the term; it starts out with fried steak and salads, cooked on an open-air grille, and served with plates, forks, and other paraphernalia; this sort of thing, like the elegance of the average American home and of the women's clothes, seems to me rather a rebirth of the Victorian era, flourishing over here by virtue of the same conditions that nourished it in England. Not only in manners, but also in politics and international affairs, I often feel that Victorian England and modern America would understand each other better than either understands its contemporaries.

I don't remember how much I have told you about my plans; they were greatly helped by an offer of a ride across the country by Feynman, the bright young professor of whom I have often spoken. He is going to visit his (Catholic) sweetheart at Albuquerque, New Mexico, and is driving across the country starting this week; I am to go to keep him company on the way out, and I shall leave him and make my own way to Ann Arbor as soon as I (or we) feel I have had enough. It should be a fine trip, and we shall have the whole world to talk about. On this visit Feynman intends to make up his mind either to marry the girl or to agree to part; most people are prepared to wager for the former alternative.

Two weeks later comes the next extract, written from Santa Fe, New Mexico.

25 June 1948

Feynman originally planned to take me out West in a leisurely style, stopping and sight-seeing en route and not driving too fast. However, I was never particularly hopeful that he would stick to this plan, with his sweetheart waiting for him in Albuquerque. As it turned out, we did the 1800 miles from Cleveland to Albuquerque in 3 1/2 days, and this in spite of some troubles; Feynman drove all the way, and he drives well, never taking risks but still keeping up an average of 65 mph outside towns. It was a most enjoyable drive, and one could see most of what was to be seen of the scenery without stopping to explore; the only regret I have is that in this way I saw less of Feynman than I might have done.

At St. Louis we joined Highway US 66, the so-called "Main Street of America" which runs from Chicago to Los Angeles via Albuquerque. We thought that from there on would be plain sailing, as this is one of the best marked and maintained roads there is. However, at the end of the second day we ran into a traffic jam, and some boys told us that there were floods over the road ahead and no way through. We retreated to a town called Vinita, where with great difficulty we found lodging for the night, the town being jammed with stranded travellers. We ended up in what Feynman called a "dive," viz. a hotel of the cheapest and most disreputable character, and with a notice posted in the corridor saying "This Hotel is under New Management, so if you're drunk you've come to the wrong place." During the night it rained continuously, and the natives said it had been raining most of the time for more than a week.

The fourth day we drove the last 300 miles to Albuquerque before 1 pm. This was the most beautiful part of the trip, though I was surprised to find how little of it was typical New Mexico mountains; the prairie actually extends half-way across New Mexico, and only the last 20 miles of our journey were in mountains, the Sandia range immediately east of the Rio Grande valley in which Albuquerque lies. As we advanced into New Mexico the prairie grew drier and drier, until a fair proportion of the vegetation was cactus, carrying at this time of year a profusion of large blood-red flowers. Coming down into Albuquerque, Feynman said he hardly recognised the place, so much has it been built up since he was there 3 years ago. It is a fine, spacious town of the usual American type; very little of the Spanish surviving.

Sailing into Albuquerque at the end of this Odyssey, we had the misfortune to be picked up for speeding; Feynman was so excited that he did not notice the speed limit signs. So our first appointment in this *romantic city of home-coming was an interview with the Justice of the Peace; he was a pleasant enough fellow, completely informal, and ended up by fining us $10 with $4.50 costs, while chatting amiably about the way the Southwest was developing. After this Feynman went off to meet his lady, and I came up by bus to Santa Fe.*

All the way, Feynman talked a great deal about this sweetheart, his wife who died at Albuquerque in 1945, and marriage in general. Also about Los Alamos, and life and philosophy in general. I came to the conclusion that he is an exceptionally well-balanced person, whose opinions are always his own and not other people's! He is very good at getting on with people, and as we came West he altered his voice and expressions unconsciously to fit his surroundings, until he was saying "I don't know noth'n' " like the rest of them.

QED

After leaving Feynman in Albuquerque, I spent the summer at Ann Arbor and Berkeley, and succeeded in understanding how the Schwinger and Feynman versions of electrodynamics fitted together. In September I arrived at the Institute for Advanced Study in Princeton. The next extract was written from Princeton, in response to my father's request for an explanation of what I had been doing. My father had called attention to a clause in the Athanasian Creed that says, "There is the Father incomprehensible, and the Son incomprehensible and the Holy Ghost incomprehensible, yet there are not three incomprehensibles but one incomprehensible." He remarked that this sounded a bit like Schwinger, Feynman and Dyson. Here is my reply:

4 October 1948

To-day came a fat letter from you. Concerning your remarks about the Athanasian Creed, I think what you say is very much to the point; but I must disappoint any hopes that Dyson-Schwinger-Feynman might prove an effective substitute. Seriously, I should like some time, now that I understand these theories properly, to try to write some intelligible semi-popular account of them; but it would be a difficult job, not to be undertaken in one afternoon. The central idea of the theories, in any case, is to give a correct account of experimental facts while deliberately ignoring certain mathematical inconsistencies which come in when you discuss things that cannot be directly measured; in this there is a close similarity to the Athanasian Creed. However, there is the

important difference that these theories are expected to last only about
as long as no fresh experiments are thought up, so they will hardly do
as a basis for a Weltanschauung.

This letter shows that in 1948 we all thought of the new quantum
electrodynamics as a ramshackle structure, soon to be replaced by
something better. We would never have believed it if anyone had told
us that our theory would still be around 40 years later, the calculated
effects of radiative reactions still agreeing with experiments to an ac-
curacy of 10 or 11 significant figures. We would have been even more
surprised to learn that the ugliest and most awkward features of our
theory, the tricky renormalization of mass and charge, would remain
as key features when the theory was eventually incorporated, first into
the Weinberg-Salam-Glashow unified theory of electromagnetic and
weak interactions, and later into the grand unified theories of weak
and strong interactions. For Feynman, renormalization was not something to be proud of but something to be got rid of. He spent a lot of
time trying to construct a finite-electron theory that would make
renormalization unnecessary.

Cécile Morette

In the next letter a great woman appears, whose name was then
Cécile Morette and is now Cécile DeWitt. In 1948, she was a member of
the Institute for Advanced Study, having arrived from France via Dub-
lin and Copenhagen. She was the first of the younger generation to
grasp the full scope and power of the Feynman path integral approach
to physics. While I was concerned with applying Feynman's methods
to detailed calculations, she was thinking of larger issues, extending
the path integral idea to everything in the universe including gravitation and curved space-times.

Boston
1 November 1948

After my last letter to you I decided that what I needed was a long
week-end away from Princeton, and so I persuaded Cécile Morette to
come with me to see Feynman at Ithaca. This was a bold step on my
part, but it could not have been more successful and the week-end was
just deliriously happy. Feynman himself came to meet us at the station,
after our 10-hour train journey and was in tremendous form, bubbling

over with ideas and stories and entertaining us with performances on Indian drums from New Mexico until 1 am.

The next day, Saturday, we spent in conclave discussing physics. Feynman gave a masterly account of his theory, which kept Cécile in fits of laughter and made my talk at Princeton a pale shadow by comparison. He said he had given his copy of my paper to a graduate student to read, then asked the student if he himself ought to read it. The student said "No" and Feynman accordingly wasted no time on it and continued chasing his own ideas. Feynman and I really understand each other; I know that he is the one person in the world who has nothing to learn from what I have written; and he doesn't mind telling me so. That afternoon, Feynman produced more brilliant ideas per square minute than I have ever seen anywhere before or since.

In the evening I mentioned that there were just two problems for which the finiteness of the theory remained to be established; both problems are well-known and feared by physicists, since many long and difficult papers running to 50 pages and more have been written about them, trying unsuccessfully to make the older theories give sensible answers to them. Amongst others, [Nicholas] Kemmer and the great [Werner] Heisenberg had been baffled by these problems.

When I mentioned this fact, Feynman said, "We'll see about this," and proceeded to sit down and in two hours, before our eyes, obtain finite and sensible answers to both problems. It was the most amazing piece of lightning calculation I have ever witnessed, and the results prove, apart from some unforeseen complication, the consistency of the whole theory.

The two problems were, the scattering of light by an electric field, and the scattering of light by light.

After supper Feynman was working until 3 am. He has had a complete summer of vacation and has returned with unbelievable stores of suppressed energy.

On the Sunday Feynman was up at his usual hour (9 am) and we went down to the physics building, where he gave me another 2-hour lecture on miscellaneous discoveries of his. One of these was a deduction of [James] Clerk Maxwell's equations of the electromagnetic field from the basic principles of quantum theory, a thing which baffles everybody including Feynman, because it ought not to be possible.

At 12 on the Sunday we started our journey home, arriving finally at 2 am and thoroughly refreshed. Cécile assured me she had enjoyed it as much as I had.

Feynman's proof of the Maxwell equations is still unpublished and still after 40 years as baffling as ever. He refused to publish it then because he claimed it was only a joke. I am not so sure. The proof is mathematically rigorous and does exactly what I said in the letter. He starts with the commutation relations between position and velocity of a single nonrelativistic particle obeying Newton's law of motion and deduces the existence of electric and magnetic fields satisfying the equations of Maxwell. Expressed in modern language, the proof shows that the only possible fields that can consistently act on a quantum mechanical particle are gauge fields. I am sorry I never succeeded in persuading Feynman to publish it.

The next letter gives us another glimpse of Cécile Morette.

14 November 1948

Cécile amused us all yesterday by bringing down a French millionaire to see the institute (an industrial magnate of some kind). She said she hinted to him fairly strongly that France could do with an institute of a similar sort; she said if she were made director of the French institute she would invite all of us to come and lecture there. It will be interesting to see if anything comes of it.

The man Cécile brought to see the Princeton institute was Léon Motchane. Motchane later became the founder and first director of the Institut des Hautes Etudes Scientifiques at Bures-sur-Yvette in France. The IHES is a flourishing institution that has made a major contribution to the support of mathematics and theoretical physics in France. Cécile was 26 when she brought Motchane to Princeton and planted the seed that grew into IHES. A few years later she founded the Les Houches summer school, which is also a flourishing institution and has been a training ground for generations of European students.

My last extract from the year 1948–49 describes a visit by Feynman to the institute at Princeton. Those who have read "Surely You're Joking, Mr. Feynman!" will know that Feynman hated the institute. J. Robert Oppenheimer offered him an institute professorship and Feynman turned it down flat. Feynman considered the institute to be snobbish, stuffy and scientifically sterile. He was invited many times to visit but he almost never came. He did come once, as the next extract records.

Chicago
28 February 1949

On Thursday we had Feynman down to Princeton, and he stayed till I left on Sunday. He gave in 3 days about 8 hours of seminars, besides long private discussions. This was a magnificent effort, and I believe all the people at the institute began to understand what he is doing. I at least learnt a great deal. He was as usual in enthusiastic mood, waving his arms about a lot and making everybody laugh. Even Oppenheimer began to get the spirit of the thing, and said some things less sceptical than is his habit. Feynman was obviously anxious to talk, and would have gone on quite indefinitely if he had been allowed; he must have been suffering from the same bottled-up feeling that I had when I was full of ideas last autumn. The trouble with him, of course, is that he never will publish what he does; I sometimes feel a little guilty for having cut in in front of him with his own ideas. However, he is now at last writing up two big papers, which will display his genius to the world.

A Letter from an Infrequent Correspondent

That is the end of my meager record of the young Feynman. The next item in my file is the one and only letter written by Feynman to me. Since Feynman letters are even rarer than Feynman papers, I am glad to put this one on record.

Cornell University
15 May 1950

Dear Freeman: I have heard that Cornell University is offering you a position. I was very glad to hear that because I think you would like it here very much. Of course you have been here a year and know how good it is to work with Bethe, etc., so there is no point in going into all these things, but I did want to tell you that the administration of the physics department of Cornell is excellent. It is completely free of politics and petty quarrels, etc., which makes it an ideal place to work. If a man wants to do mostly research in preference to teaching or vice versa either way is arranged. I have been very happy here.

My main point in writing to you is to assure you that I did not leave Cornell for any reason that had to do with the failure of the university, it is just that I have no family or anything to tie me down and the

wanderlust set in, so I thought I would like to live for a while in Cali-
fornia, that is all there is to it. I hope my doing this has gotten you a real
good job.

Sincerely yours,
R. P. Feynman [sic]

Knowing how much it cost Feynman to write a letter, I was deeply
grateful for his encouragement. I went back to Cornell in the fall of
1951, after he had moved to Caltech. Cornell treated me well, but without Feynman the place seemed sad and empty. The next encounter
with Feynman in my file is 29 years later.

Princeton
21 December 1979

The best thing that happened was a supper with Dick Feynman at his
home in Pasadena. The first time we had met for about 12 years. He had
in the meantime had a big cancer operation 1 1/2 years ago. Rumor had
said he was dying but I found him in bouncing health and spirits. He is
still the same old Feynman that drove with me to Albuquerque 30 years
ago. He said they took a 6-pound lump out of him but since it was all
fatty tissue (liposarcoma) there is a good chance it was contained and
will not recur.

Feynman has been married for about 20 years to an English wife
called Gweneth. He enjoys the domestic life and they have a menagerie
very much like ours, 1 horse (for the 12-year-old daughter), 2 dogs, 1
cat, 5 rabbits. But they have temporarily outdone us, for the next few
months, by taking on a boa constrictor who belongs to some neighbours
on a leave of absence.

Another editorial comment here. I heard later from Feynman that this
affair with the boa constrictor ended badly. Feynman was responsible
for the care and feeding of the boa constrictor. The trouble was that the
boa constrictor was supposed to eat live white mice, but when Feynman fed him the mice he was too stupid or too lazy to catch them.
Instead of the boa constrictor eating the mice, the mice began eating
the boa constrictor. So Feynman had to sit up at night to stop the mice
from nibbling holes in the boa constrictor's skin. And then, when the
owner of the boa constrictor came back, she scolded Feynman for
taking care of the animal so badly. He said he had learned something
from this experience. In the future, if he was asked to baby-sit for a boa
constrictor, he would say no. It was a no-win situation. He said it's all

right to say yes once to a dumb idea, to try it out and see if it is any good. But you are a fool if you say yes twice to the same dumb idea. After things turn out badly the first time, you say no.

Feynman in Later Years, Unchanged

Now comes the last letter in my collection, written to Sara Courant. Sara had been at Cornell in 1948 when her husband Ernest was a postdoc.

Urbana, Illinois
9 April 1981

Dear Sara:

I just spent a marvelous three days with Dick Feynman and wished you had been there to share him with us. Sixty years and a big cancer operation have not blunted him. He is still the same Feynman that we knew in the old days at Cornell.

We were together at a small meeting of physicists organized by John Wheeler at the University of Texas. For some reason Wheeler decided to hold the meeting at a grotesque place called World of Tennis, a country club where Texas oil millionaires go to relax. So there we were. We all grumbled at the high prices and the extravagant ugliness of our rooms. But there was nowhere else to go. Or so we thought. But Dick thought otherwise. Dick just said: "To hell with it. I am not going to sleep in this place," picked up his suitcase and walked off alone into the woods. In the morning he reappeared, looking none the worse for his night under the stars. He said he did not sleep much, but it was worth it.

We had many conversations about science and history, just like in the old days. But now he had something new to talk about, his children. He said: "I always thought I would be a specially good father because I wouldn't try to push my kids into any particular direction. I wouldn't try to turn them into scientists or intellectuals if they didn't want it. I would be just as happy with them if they decided to be truck-drivers or guitar-players. In fact I would even like it better if they went out in the world and did something real instead of being professors like me. But they always find a way to hit back at you. My boy Carl for instance. There he is in his second year at MIT, and all he wants to do with his life is to become a god-damned philosopher!"

As we sat in the airport waiting for our planes, Dick pulled out a pad of paper and started to draw faces of people sitting in the lounge. He drew them amazingly well. I said I was sorry I have no talent for drawing. He said: "I always thought I have no talent either. But you don't need any talent to do stuff like this. Just a few years ago I made friends with an artist and we had an agreement. I would teach him quantum mechanics and he would teach me drawing. We both did quite well, but he was a better teacher than I was."

So I left him there in the airport at Austin, Texas, happily talking about Michelangelo and Raphael and Giotto with all the enthusiasm of a teenager. He said, "You know, I went in there to look at the Sistine Chapel in Rome, and I could see at once that one of those panels wasn't as good as the others. It just wasn't good. And then afterwards I looked in the guide-book and it said that particular one was painted by somebody else. That made me feel good. I always thought Art Appreciation was a lot of hokum, but now I found out I can do it myself."

That was not the last time I saw Feynman, but it is the last glimpse of him recorded in my letters. Fortunately, we have a much more complete picture of Feynman in his own words, recorded by Ralph Leighton and published in the two books *"Surely You're Joking, Mr. Feynman!"* and *"What Do You Care What Other People Think?"* If you look carefully at the chapter "Hotel City" in the second book, you will find Feynman's version of the night he and I spent together in the hotel in Vinita in 1948. His version is very different from the version I wrote in my letter to my parents. In deference to my parents' Victorian sensibilities, I left out the best part of the story. Feynman always remembered what his first wife, Arline, had written to him from her hospital bed before he went to Los Alamos: "What do you care what other people think?" Her spirit stayed with him all his life and helped to make him what he was, a great scientist and a great human being.

This article was originally published in the February 1989 issue of *Physics Today*.

TO HAVE BEEN A STUDENT OF RICHARD FEYNMAN

Laurie M. Brown

It was my good fortune to study physics as a graduate student at Cornell University during 1946–50 when Richard Feynman was on the faculty. I attended his courses in electrodynamics, mathematical methods of physics, quantum mechanics, elementary and advanced, and took part in the theoretical physics seminar. In the latter there were spirited discussions involving Feynman and the other elementary particle and field theorists at Cornell. These included Professors Hans Bethe and Philip Morrison, and postdocs Edwin Salpeter, who became a well-known astrophysicist, Ning Hu, who had worked with Wolfgang Pauli (and who soon returned to China where he played an important role after the Communist revolution), Fritz Rohrlich, and Robert Gluckstern. Among the graduate students was the very young-looking and prodigious Freeman Dyson. The theorists also participated actively at the experimental seminars, notably those on cosmic rays and high energy physics, with the experimentalists led by Robert R. (Bob) Wilson, the future founding director of Fermilab, and including Robert Bacher, Boyce McDaniel, Kenneth Greisen, Dale Corson, John DeWire, William Woodward, and Giuseppe and Vana Cocconi.

Laurie M. Brown is Professor of Physics and Astronomy at Northwestern University, Illinois.

In 1948 I became Hans Bethe's assistant, and soon afterwards persuaded Feynman to become my thesis advisor. After the war, government funding for research at universities became widespread, but academic scientists were not sure how to handle it without incurring criticism, either from without or within the university. At Cornell, the Laboratory of Nuclear Studies, where the nuclear, cosmic ray, and high energy physics groups were located, was largely supported by the Office of Naval Research. The faculty decided that since the money was explicitly awarded for the support of research, not education, graduate students should not be paid for working on their own theses. (This policy was changed in a few years).

Feynman had no paid assistant and, in any case, if I worked with him on my thesis, I could not be *his* research assistant, and so I had to be content with Bethe. (Of course, being able to work for both of these great physicists was an unbelievable stroke of luck for me!) Bethe had many thesis students at a time, perhaps as many as eight, as he seemed to have an inexhaustible supply of problems, but Feynman never had more than two students. Throughout his career, at Cornell and later at Cal Tech, he accepted few doctoral students. In part, as he explained, that was because he liked to concentrate on one, or at most two problems at any given time, and because he would assign only those problems for which he genuinely wished to know the answer.

Feynman was a very popular teacher, and his advanced lecture courses were well attended by both theorists and experimentalists. The students met in small groups afterwards to compare their lecture notes and to work through them again. Although the lectures were thoughtfully prepared, they were not always easy to follow. We soon learned that Feynman's methods were anything but trivial, not to be found in any books, and that some of his views were highly unconventional. Feynman stressed creativity—which to him meant working things out from the beginning. He urged each of us to create his or her own universe of ideas, so that our products, even if only answers to assigned classwork problems, would have their own original character—just as his own work carried the unique stamp of *his* personality. Obviously, that kind of teaching extends well beyond physics, or even science in general. It was excitingly different from what most of us had been taught earlier.

Some of us, unfortunately, also learned something that could handicap us for years to come: Feynman urged us not to read very widely, but to try to work out everything by ourselves from first principles. That method could serve an outstanding mind, already well-prepared, like that of Feynman, but while it inspired us to try for originality after we left Cornell, it also lowered our productivity to a point that at times was dangerous to our academic careers. In truth, a good deal of Dick's supposed naiveté, his apparent ignorance of book learning, his disdain for mathematical rigor, etc., was feigned (although his poor pronunciation and spelling was not). It was all in the spirit of his *"Surely You're Joking ..."* stories—but he conveyed it seriously enough to influence some of his admiring students, including me. (Working as Bethe's assistant did not encourage a lot of reading either, since Bethe seemed to carry all the essential physics in his head, and was able to calculate quickly any result that he wanted.) Feynman's informality was, in part, an act put on to command attention and to impress his young audience. On the whole, however, he was a careful, responsive, and caring teacher and we students felt that he was serving as our guide, not our judge.

In 1947–48, I had two semesters of quantum mechanics, first with Feynman and then with Bethe; the following year, I took Feynman's course in advanced quantum mechanics. At that time, he was working actively, indeed one might say compulsively, on his diagrammatic approach to renormalized quantum electrodynamics (QED), the work for which he shared the Nobel Prize in 1965. His course made use of the new methods that he was developing, and we students became expert in the use of Feynman diagrams before they were written up for publication. Some of Bethe's doctoral students used the Feynman diagram methods to do calculations based upon the weak-coupling meson theories that were very popular then. Unfortunately, the meson theory results turned out to have little relevance to nature, or to future developments of theoretical physics.

Since Feynman was only twenty-eight years old when he came to Cornell, and because many of the graduate students' studies had been interrupted by war service, it turned out that most of the graduate students were only a few years younger than Feynman. Some of them had served at Los Alamos during the war and were on familiar terms with Cornell faculty members who had worked there, including Feynman, Bethe, Morrison, Wilson, and others. Socially, things tended to be informal, partly because of the relatively small age spread, and partly as a carryover from the spirit of wartime collaboration, which had created a kind of meritocracy and a disregard of seniority, although to some extent the Los Alamos pecking order was maintained.

Some of the theoretical students resuming their studies had learned earlier what came to be called (after the "manifestly covariant" version became established in the late 1940s) the "old-fashioned" field theory. This formulation of quantum electrodynamics was due mainly to Heisenberg and Pauli, and its applications to the meson theory of nuclear forces were to be found in Gregor Wentzel's book, *The Quantum Theory of Fields*. Feynman was always looking for a contest of some sort, and he liked to challenge the more knowledgeable students with this sort of query: "According to Pauli, how many virtual photon polarizations must be exchanged?" Then after receiving the conventional answer (namely, two), he would say, "But that's wrong, it should be four because ..." A student working on a thesis that made use of the accepted field theory methods would develop anxiety, becoming upset at being asked to carry the burden of defending a theory for which he was not really responsible. (I know I should say "he or she"—but there were no female theory graduate students, although there were several excellent female experimenters.) Their concern was this: If this adventurer, Feynman, bent on destroying conventional field theory, were to be successful, might the student's thesis turn out to be wrong, and possibly not be accepted?

On the other hand, students who took Feynman's courses and learned his methods tended to disregard the Heisenberg-Pauli methods—and indeed had a totally unjustified contempt for them that would have to be unlearned later. That was one example of Feynman's intensity being able to carry everything before him. He drove himself, working literally day and night. When he felt challenged to prove a conjecture or to complete a calculation, he would work all through the night. Then, for some weeks he might do no physics at all. Of his thesis students he demanded a comparable intensity, and he assigned them problems that were extensions of his own research and for which he was eager for results. Consequently he had few students, for usually he was concentrating on a single problem. One very good student changed from being a theoretical to an experimental physicist in an instant, when Feynman came in one Monday and told him that during the weekend he had solved the thesis problem that Feynman had assigned to him six months earlier.

At Cornell, Feynman was noted for unusual social behavior, the sort featured in his popular sketches, *"Surely You're Joking, Mr. Feynman."* For example, he liked to attend dances for undergraduates, where he was usually thought to be a student, based upon his youthful appearance and exuberance. However, his devil-may-care pose may have been a concealed reaction to grief and loneliness. Readers of the second of the two autobiographical collections, *"What do you care what other people think?,"* will recall from his moving memoir on the subject that Feynman's first wife, Arline, had died of a lingering illness in 1946, while he was at Los Alamos. In spite of the extroverted *persona* that he liked to present, he was actually a very private person, who reacted angrily whenever anyone tried to penetrate this facade, as I have observed faculty wives and others try. Once he was persuaded by a psychology graduate student to take a Rohrschach test, but during the test he would discuss only the physical production of the inkblot. For example, he would say, "This one was made with thumb and middle finger pressed together," and so forth. Afterwards, he told me about it, stating proudly that he had outwitted the analyst!

During the time that I was his doctoral student, he was writing his famous papers on the theory of positrons (with the backward-in-time paths) and on QED, which appeared together in the same issue of the Physical Review. He asked me to read them, to correct his terrible spelling and grammar, and to serve as a guinea pig. Anything that was not clear to me, he said he would rewrite, as he wanted it to be readable by someone at my level of ability and experience. Thus I had a marvelous opportunity to discuss these important works with him at length. He assigned to me at first a thesis topic that bore fruit only in later years, but which allowed me to have a taste of being creative in physics. (However, my actual thesis topic turned out to be a different one.)

Pauli had noted that a term could be added to Dirac's relativistic Hamiltonian for the electron that would represent an anomalous magnetic moment interaction without violating any accepted invariance requirements. This could, it appeared, be used to represent the neutron's electromagnetic interaction, for example. Feynman noted that a term of the same form as Pauli's appears automatically upon iterating the Dirac equation. [More explicitly, if the Dirac equation is written $(D-m)\Phi=0$, the iterated equation reads $(D+m)(D-m)\Phi=0$.] Assume now that the new Pauli term has a variable coefficient σ: For $\sigma=1$, the resulting magnetic moment is exactly the Dirac moment, for $\sigma=0$, it is zero. Feynman asked me to apply the "Feynman rules," i.e., his diagrammatic algorithm, to calculate various well-known processes for arbitrary σ. I found interesting and unexpected results: some differential "cross sections" had negative regions, although for $\sigma=1$ the Dirac results and for $\sigma=0$ the usual spin 0 results were obtained. Most interestingly, the equation had too many solutions, but the physical ones (of positive mass) could be selected if the Dirac wave function were restricted to have only two components, rather than the conventional four components.

Feynman was excited about the possibility of a two-component Dirac theory and he reported my results to Bethe. However, the latter persuaded him that the topic was unsuitable for a thesis, being too speculative and "original." Bethe felt that a standard calculation was more appropriate for an apprentice theorist. (I heard the discussion through the wall of my room, which adjoined Bethe's office. Feynman, as usual, spoke very loudly.) Accordingly, the next day I began work on the radiative corrections to the Klein-Nishina formula for Compton scattering, which became my actual doctoral thesis and which I later published with Feynman as coauthor.

The two-component Dirac equation, as I discovered later, had aleady been considered by Hendrik Kramers (and in its massless form by Hermann Weyl). Although the QED based upon it conserved reflection symmetry, the equation itself was not *manifestly* parity invariant, and it had been rejected by Pauli on those grounds. In 1957, when Feynman was working on a parity-nonconserving theory of the weak interactions, I met him at that year's Rochester Conference on High Energy Nuclear Physics, and he asked me to send to him all my notes on the two-component theory. I did so, and also wrote up the two-component version of QED as a paper, for which I asked Feynman to be a coauthor. He declined, saying that he thought I deserved to be the sole author, and so I published it alone in the *Physical Review*.

After that I saw Feynman infrequently, as he did not travel very much, but I continued to follow his splendid career with great interest. Whenever I wrote anything, and whenever I had to make a scientific judgment or decision, I always asked myself, consciously or unconsciously, "What would Feynman think of this?"—and then I tried to decide accordingly. There is no doubt in my mind that Feynman's spirit and approach to physics lives on in my mind, as it does in many other minds.

The front steps of Palmer Physical Laboratory, where Feynman had his physics courses and got banished from the cyclotron. The statue on the left depicts Ben Franklin performing his famous kite-and-key experiment. The one on the right depicts Joseph Henry, discoverer of electromagnetic self-induction and professor of natural philosophy from 1832 to 1846, before he went to Washington to be first head of the Smithsonian Institution, America's first "National Science Foundation." (Photo: Princeton University Libraries.)

Assembled in the front row of a lecture hall at the Rochester Conference of January 1952 are (front row, from left) Pierre Noyes, Freeman J. Dyson, Jack Steinberger, Richard Feynman and Hans Bethe. (Photo: AIP Emilio Segré Visual Archives.)

Gathered around a coffee table at the 1947 Shelter Island Conference are (standing, from left) Willis Lamb, John Wheeler, and (seated, from left) Abraham Pais, Richard Feynman, Herman Feshbach and Julian Schwinger. The 1947 conference is regarded as the birthplace of quantum electrodynamics. (Photo: AIP Emilio Segré Visual Archives.)

Paul A. M. Dirac and Feynman in conversation during the International Conference on Relativistic Theories of Gravitation held in Warsaw, Poland, on 25–31 July 1962. A 1933 paper by Dirac gave Feynman the key to developing a quantum version of the classical theory of electrodynamics he had worked out with Wheeler.

Shelter Island conference participants, June 1947. From left to right: I. I. Rabi, Linus Pauling, John Van Vleck, Willis Lamb, Gregory Breit, Duncan MacInnes, Karl Darrow, George Uhlenbeck, Julian Schwinger, Edward Teller, Bruno Rossi, Arnold Nordsieck, John von Neumann, John Wheeler, Hans Bethe, Robert Serber, Robert Marshak, Abraham Pais, J. Robert Oppenheimer, David Bohm, Feynman, Victor Weisskopf, Herman Feshbach. (Photo: AIP Emilio Segré Visual Archives.)

Receiving the Nobel Prize from King Gustav VI Adolf of Sweden, 10 December 1965. (Photo: California Institute of Technology.)

Richard Feynman discussing the parton model at CERN, January 1970. (Photos: CERN.)

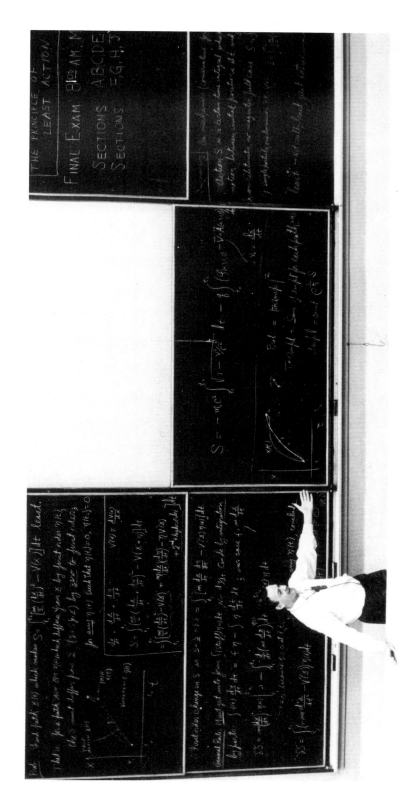

Lecturing to undergraduates on the principal of least action in early 1960s.
(Photo: AIP Emilio Segré Visual Archives.)

In the 'Court of the Oak' at Caltech in 1959, Murray Gell-Mann and Feynman enjoy a relaxed moment. (Photo: California Institute of Technology.)

Performing at a Caltech talent show, spring 1966. (Photo: California Institute of Technology.)

Talking to students in a coffee room at Caltech, June 1964. (Photos: California Institute of Technology.)

In academic robes at Caltech's commencement ceremony, June 1974. (Photo: California Institute of Technology.)

(l to r) Michiji Konuma (late President of Japanese Physical Society and Vice-President of Asian Physical Society), Chen Ning Yang, Richard and Gweneth Feynman, Laurie Brown August 1985. (Photo: Laurie Brown.)

A PATH TO QUANTUM ELECTRODYNAMICS

Julian Schwinger

On 10 December 1965 three people shared a Nobel Prize "for their fundamental work in quantum electrodynamics." I am the sole survivor of that trio. Almost a decade ago I delivered a memorial lecture for Sin-itiro Tomonaga. Now I join with others in a tribute to Richard P. Feynman.

I have been asked to write on Feynman's contribution to the development of quantum electrodynamics. In the course of the past 40 years I have had several occasions to present the history of quantum electrodynamics, and these presentations naturally included accounts of Feynman's work. But all these articles were dominated by my point of view—they were in my voice. It is more fitting here that Feynman's voice be heard. And I believe we should have him speak not about technical details, but about motives, insights and lessons for the future. The many quotations from Feynman that follow come from the three sources listed at the end of the article. I have heavily favored the Nobel lecture,[1] not only for its extensive coverage, but in the belief that, in

Julian Schwinger is University Professor Emeritus at the University of California at Los Angeles.

contrast with the other two books, it has not undergone editing, and therefore more truly projects the voice of Richard Feynman.

The Challenge

I first met Feynman at Los Alamos, about a week after the Trinity test that ushered in the age of nuclear terror. No, I was not a member of the Manhattan Project, although I did spend a little time at the Metallurgical Laboratory in Chicago to see if I wanted to join up. I didn't. I was at Los Alamos on a purely cultural mission, from the MIT Radiation Laboratory, to give a few lectures about electromagnetic waveguides and electron accelerators. The talk on the latter topic included a discussion of synchrotron radiation.

One evening I ran into Feynman, looking rather glum (perhaps Robert R. Wilson had just said to him, "It's a terrible thing that we made"?[2]). He began to lament the loss of irreplaceable time to do physics, of which I was also keenly aware; we were both 27 years old then. He said something like, "I haven't done anything, but you've already got your name on something." I still wonder what he was referring to.

It wasn't true that he hadn't done anything. Already, as an undergraduate at MIT in the late 1930s, he had realized "that the fundamental problem of the day was that the quantum theory of electricity and magnetism was not completely satisfactory."[1] Concerning the books of Walter Heitler and Paul Dirac, for example, Feynman said[1]:

> I was inspired by the remarks in those books; not by the parts in which everything was proved and demonstrated [but by] the remarks about the fact that this doesn't make any sense, and the last sentence of the book of Dirac I can still remember, "It seems that some essentially new physical ideas are here needed." So I had this as a challenge and an inspiration. I also had a personal feeling, that since they didn't get a satisfactory answer to the problem I wanted to solve, I didn't have to pay a lot of attention to what they did do.

However, Feynman did gather from his reading that[1]

> two things were the source of the difficulties with quantum electrodynamical theories. The first was an infinite

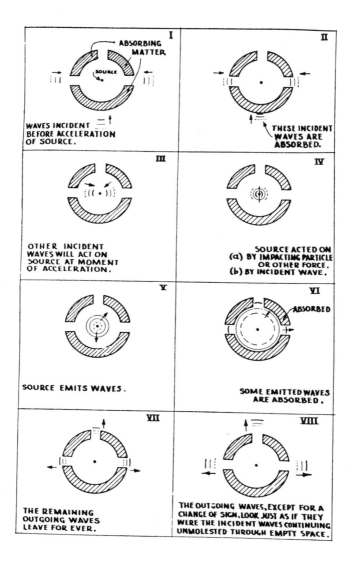

Diagrams that John Wheeler and Richard Feynman used to illustrate the constructive and destructive interactions between advanced and retarded fields in their time-symmetric electrodynamics. (From J. A. Wheeler, R. P. Feynman, Rev. Mod. Phys. *17, 157, 1945.)*

energy of interaction of the electron with itself. And this difficulty existed even in the classical theory. Well, it seemed to me quite evident that the idea that a particle acts on itself, that the electrical force acts on the same particle that generates it, is not a necessary one—it is a sort of silly one, as a matter of fact. And, so I suggested to myself, that electrons cannot act on themselves, they can only act on other electrons. That means there is no field at all.

Feynman was very happy with this. He saw it as solving what he then thought to be the second problem of quantum electrodynamics as well: the infinite vacuum energy associated with the infinite number of degrees of freedom of the electromagnetic field. No field, no infinite number of degrees of freedom. As Feynman said[1]:

That was the beginning, and the idea seemed so obvious to me that I fell deeply in love with it ... . I was held to this theory ... by my youthful enthusiasm.

Then I went to graduate school and somewhere along the line I learned what was wrong with the idea that an electron does not act on itself. When you accelerate an electron it radiates energy and you have to do extra work to account for that energy. The extra force against which this work is done is called the force of radiation resistance. The origin of this extra force was identified in those days ... as the action of the electron on itself. The first term of this action ... gave a kind of inertia. But that inertia-like term was infinite for a point charge. Yet the next term in the sequence gave an energy loss rate, which for a point-charge agrees exactly with the rate that you get by calculating how much energy is radiated. So, the force of radiation resistance, which is absolutely necessary for the conservation of energy would disappear if I said that a charge could not act on itself.

So, I learned ... the glaringly obvious fault of my own theory. But, I was still in love with the original theory, and was still thinking that with it lay the solution to the difficulties of quantum electrodynamics.

Feynman eventually took his problem to John Wheeler, for whom he was working as a research assistant during 1940–41. They came up with an answer that had two elements. The ordinary classical theory says that the motion of a charged particle at a certain time is influenced by the behavior of other charges at earlier times such that light can cover the relevant distance in the time available. Wheeler and Feynman changed this so-called retarded action-at-a-distance electrodynamics into one that is half retarded, half advanced. This, of course, seemed to wreak havoc with conventional ideas about causality. Nevertheless, it was equivalent to the retarded description and contained the radiative resistance force, provided one assumed that any emitted radiation was totally absorbed within the complete system of charges.

Wheeler and Feynman also found that, unlike the situation with retarded interactions, a theory that is symmetrical between retarded and advanced interactions permits an action principle description. That in itself was hardly new—Adriaan D. Fokker showed it in 1929, for example—although the suggestion that suitable boundary conditions could encompass the causal, dissipative situation of radiating, interacting charges certainly was.

About this success Feynman said[1]:

> I was now convinced that since we had solved the problem of classical electrodynamics (and completely in accordance with my program from MIT, only direct interactions between particles, in a way that made fields unnecessary) that everything was definitely going to be all right. I was convinced that all I had to do was make a quantum theory analogous to the classical one and everything would be solved.

Wheeler urged Feynman to give a seminar on their classical theory, promising at the same time that he would work out the quantum theory version himself and give a later seminar on that. Feynman has described how his inaugural technical lecture attracted such luminaries as John von Neumann, Wolfgang Pauli and Albert Einstein, and he records Pauli's accurate prediction that Wheeler would never give the promised seminar on the quantum formulation.

Feynman would find his own path to quantum mechanics. But before we enter on it, we should note some other aspects of this classical odyssey, beginning with "suggestions for interesting modifications of electrodynamics."[1]

The part of the action that describes the interaction between charged particles contains a discontinuous function δ, which is zero except when the space-time locations of the charges are such that they can exchange light signals—that is, when each is on the other's light cone. So, Feynman and Wheeler figured, one might

> replace this delta function ... by another function, say, f, which is not infinitely sharp [but is] a narrow peaked thing, [and] all of the tests of electrodynamics that were available in Maxwell's time would be adequately satisfied ... . You have no clue of precisely what function to put in for f, but it was an interesting possibility to keep in mind when developing quantum electrodynamics.

> It also occurred to us that if we did that (replace δ by f) we could reinstate [the term referring to a single charge] ... a finite action of a charge on itself. In fact, it was possible to prove that ... the main effect of self-action ... would be to produce a modification of the mass. [Indeed] all the mechanical mass could be electromagnetic self-action.

Looking back at this part of his voyage, Feynman said[1]:

> I would also like to emphasize that by this time I was becoming used to a physical point of view different from the more customary [view, in which] things are discussed as a function of time in very great detail. For example, you have the field at this moment, a differential equation gives you the field at the next moment and so on; a method which I shall call the Hamiltonian method ... . We have, instead (in [the action]) a thing that describes the character of the path throughout all of space and time.

Dirac's Groundwork

In 1933 Dirac published a paper in *Physikalische Zeitschrift der Sowjetunion* on "The Lagrangian in Quantum Mechanics." He begins by saying:

> Quantum mechanics was built up on a foundation of analogy with the Hamiltonian theory of classical mechanics.

This is because the classical notion of canonical coordinates and momenta was found to be one with a very simple quantum analogue ... .

Now there is an alternative formulation for classical dynamics, provided by the Lagrangian. This requires one to work in terms of coordinates and velocities instead of coordinates and momenta. The two formulations are, of course, closely related, but there are reasons for believing that the Lagrangian one is the more fundamental. In the first place the Lagrangian method allows one to collect together all the equations of motion and express them as the stationary property of a certain action function. (This action function is just the time integral of the Lagrangian.) There is no corresponding action principle in terms of the coordinates and momenta of the Hamiltonian theory. [This is not true, but it doesn't matter.] Secondly the Lagrangian method can easily be expressed relativistically, on account of the action function being a relativistic invariant; while the Hamiltonian method is essentially nonrelativistic in form, since it marks out a particular time variable ... .

For these reasons it would seem desirable to take up the question of what corresponds in the quantum theory to the Lagrangian method of the classical theory.

From the earliest days of nonrelativistic wave mechanics it had been recognized that the expression of a wavefunction as $\exp[(i/\hbar)W]$, with W expanded in powers of $\hbar$, gave the semiclassical approximation, in which the leading term of W, the one independent of $\hbar$, is the classical action. The next term, imaginary and proportional to $\hbar$, can then be found by integration over known classical quantities. Furthermore, for a free particle these two terms suffice to give the exact answer. That is, they completely determine, respectively, the real phase and real amplitude of the free-particle wavefunction. (It might be added that a particular form of the free-particle wavefunction was well known through the evident analogy between the Schrödinger equation and the heat conduction or diffusion equations.)

Dirac considered the wavefunction that relates a coordinate eigenvalue state at one time, say t_1, to any analogous state at another time, say t_2. The totality of such wavefunctions for fixed times t_1 and t_2

constitutes the time transformation function that connects the descriptions of the physical system at the two times. As the inventor of quantum transformation theory, Dirac knew, and stated explicitly, that the transformation function in question could be constructed from a sequence of transformation functions relating states at times intermediate between t_1 and t_2. In the limit where these successive intermediate times differ infinitesimally, the transformation function appears as an infinity of independent integrals extended over all coordinate values, each integral being labeled by a value of the time between t_1 and t_2. The integrand is the product of all the transformation functions associated with the successive infinitesimal increments of time.

And what is the transformation function associated with the infinitesimal displacement from time t to time $t+dt$? Dirac says it *corresponds* to $\exp[(i/\hbar)dt\, L]$, where "we ought to consider the classical Lagrangian, not as a function of the coordinates and velocities, but rather as a function of the coordinates at time t and the coordinates at time $t+dt$." Then the integrand is $\exp[(i/\hbar)W]$, where

$$W = \int_{t_2}^{t_1} dt\, L$$

This integral, which Dirac denotes by F, is the sum over all the individual coordinate-dependent terms that refer to the successive values of t.

Now, we know, and Dirac surely knew, that to within a constant factor the "correspondence," for infinitesimal dt, is an equality when we deal with a system of nonrelativistic particles possessing a coordinate-dependent potential energy V. (Relative to noninteracting particles, the presence of V only supplies an additional phase factor, that which is conveyed by the contribution $-V$ in L.) Why, then, did Dirac not make a more precise, if less general, statement? Because he was interested only in a general question: What, in quantum mechanics, corresponds to the classical principle of stationary action?

Dirac answered his fundamental question with the aid of the formal device that represents the classical limit as the limit $\hbar \to 0$. Evidently in that limit $\exp[(i/\hbar)W]$ will in general have an infinitely oscillatory dependence on any of its myriad of variables. Thus the multiple-integral construction of the transformation function "contains the quantum analogue of the action principle, [because] the importance of our considering any set of values for the intermediate [coordinates] is determined by the importance of this set of values in the integration. If we now make $\hbar$ tend to zero, this statement goes over into the classical statement that ... the importance of our considering any set of values for the intermediate [coordinates] is zero unless these values make the action function stationary."

Path Integral Formulation

Why, in the decade that followed, didn't someone pick up the computational possibilities offered by this integral approach to the time transformation function? To answer this question bluntly, perhaps no one needed it—until Feynman came along. He has described how, at a Princeton beer party, he was accosted by Herbert Jehle, newly arrived from Europe, who wanted to know what Feynman was working on. After telling Jehle about his struggles with electrodynamics, Feynman turned to Jehle and asked, "Listen, do you know any way of doing quantum mechanics starting with action?"[1] As it happened, Jehle was aware of Dirac's early paper, and so Feynman found what he wanted, a formulation of quantum mechanics that could be applied to his classical action-at-a-distance electrodynamics—if one took for granted that Dirac's construction still worked when a Lagrangian did not exist. Feynman called this approach to quantum mechanics the path integral formulation because a value of the action W is assigned to any sequence of intermediate coordinate values—to any path between the initial and the final coordinates—and all such values of $\exp[(i/\hbar)W]$ are added together.

It didn't take Feynman long to discover that[1]

> I could not get the thing to work with the relativistic case of spin one-half. However, although I could deal with the matter [electrons] only nonrelativistically, I could deal with the light or the photon interactions perfectly well ... .

> It was also possible to discover what the old concepts of energy and momentum would mean with this generalized action. And, so I believed that I had a quantum theory of classical electrodynamics—or rather of this new classical electrodynamics described by [the half-retarded, half-advanced action] ... . It was also easy to guess how to modify the electrodynamics, if anybody ever wanted to modify it. I just changed the delta to an f, just as I would for the classical case. So, it was very easy, a simple thing ... . Yet,

as I worked out many of these things and studied different forms and different boundary conditions, I got a kind of funny feeling that things weren't exactly right. I could not clearly identify the difficulty and in one of the short periods during which I imagined I had laid it to rest, I published a thesis and received my PhD.

Feynman became involved with the Manhattan Project at an early stage. He had been recruited by Wilson to work on a method for separating isotopes of uranium, which, as it turned out, was never used. He was one of the first to arrive when the Los Alamos Laboratory began in 1943. About the war years and his preoccupation with quantum electrodynamics, Feynman said[1]:

> During the war, I didn't have time to work on these things very extensively, but wandered about on buses and so forth, with little pieces of paper, and struggled to work on it and discovered indeed that there was something wrong, something terribly wrong. I found that if one generalized the action from the nice Lagrangian forms [that is, from the time integral of the Lagrangian to the action of action-at-a-distance electrodynamics] then the quantities which I defined as energy, and so on, would be complex. The energy values of stationary states wouldn't be real and probabilities of events ... would not add up to one.

Feynman summarized the position he was in prior to the Shelter Island conference of June 1947 as follows[1]:

> I would say, I had much experience with quantum electrodynamics, at least in the knowledge of many different ways of formulating it, in terms of path integrals of actions and in other forms. One of the important by-products, for example, of much experience in these simple forms, was that it was easy to see how to combine together what was in those days called the longitudinal and transverse fields, and in general to see clearly the relativistic invariance of the theory. [But] I never used all that machinery which I had cooked up to solve a single relativistic problem. I hadn't even calculated the self-energy of an electron up to that moment, and was studying the difficulties with the

conservation of probability, and so on, without actually doing anything, except discussing the general properties of the theory.

Experimental Input

The Lamb-shift measurement, and Hans Bethe's nonrelativistic calculation that accounted for a major portion of it, spotlighted the need for an effective relativistic quantum electrodynamics. Bethe had suggested that a theory giving finite results, even if it violated some physical principle, would be useful in identifying the physical quantities of interest. Feynman was sure he knew how to do that, until he tried it and, as he said,[1]

> finally realized that I had to learn how to make a calculation. So, ultimately, I taught myself how to calculate the self-energy of an electron. [Then] I simply followed the program outlined by Professor Bethe and showed how to calculate all the various things, the scattering of electrons from atoms without radiation, the shift of levels and so forth, calculating everything in terms of the experimental mass ... .

> The rest of my work was simply to improve the techniques then available for calculations, making diagrams to help analyze perturbation theory quicker. Most of this was first worked out by guessing—you see, I didn't have the relativistic theory of matter. For example, it seemed to me obvious that the velocities in nonrelativistic formulas have to be replaced by Dirac [matrices]. I just took my guess from the forms that I worked out using path integrals for nonrelativistic matter, but relativistic light. It was easy to develop rules. In addition, I included diagrams ... improved notations ... worked out easy ways to evaluate integrals ... and made a kind of handbook on how to do quantum electrodynamics.

> But one step of importance ... involved ... the negative energy sea of Dirac, which caused me so much logical difficulty. [Here Feynman recalls a suggestion by Wheeler that

a positron is an electron going backward in time.] Therefore, in the time-dependent perturbation theory that was
usual for getting self-energy, I simply supposed that for a
while we could go backward in time. [The extra terms thus
produced] were the same as the terms that other people
got when they did the problem ... using holes in the sea,
except, possibly, for some signs. These, I, at first, determined empirically by inventing and trying some rules.

I have tried to explain that all the improvements of relativistic theory were at first more or less straightforward,
semi-empirical shenanigans. Each time I would discover
something, however, I would go back and I would check it
so many ways ... until I was absolutely convinced of the
truth of the various rules and regulations which I con-
cocted to simplify all the work ... .

At this stage, I was urged to publish this because everybody said it looks like an easy way to make calculations,
and wanted to know how to do it. I had to publish it,
missing two things; one was a proof of every statement in
a mathematically conventional sense. Often, even in a
physicist's sense, I did not have a demonstration of how to
get all of these rules and equations, from conventional
electrodynamics. But, I did know from experience, from
fooling around, that everything was, in fact, equivalent to
the regular electrodynamics and had partial proofs of
many pieces, although, I never really sat down, like Euclid
did for the geometers of Greece, and made sure that you
could get it all from a single simple set of axioms. As a
result, the work was criticized, I don't know whether fa-
vorably or unfavorably, and the "method" was called the
"intuitive method." For those who do not realize it, however, I should like to emphasize that there is a lot of work
involved in using this "intuitive method" successfully. Because no simple clear proof of the formula or idea presents itself, it is necessary to do an unusually great amount
of checking and rechecking for consistency and correct-
ness in terms of what is known ... . In the face of the lack
of direct mathematical demonstration ... one should make
a perpetual attempt to demonstrate as much of the for-

mula as possible. Nevertheless, a very great deal more truth can become known than can be proven ... .

This brings me to the second thing that was missing when I published the paper, an unresolved difficulty. With δ replaced by f the calculations would give results ... for which the sum of the probabilities of all alternatives was not unity ... . I believe there is really no satisfactory quantum electrodynamics, but I'm not sure.

Therefore I think that the renormalization theory is simply a way to sweep the difficulties of the divergences of electrodynamics under the rug. I am, of course, not sure of that.

Some 20 years later Feynman had not really changed his mind, writing that[3]

The shell game that we play [is] called "renormalization." But no matter how clever the word, it is what I would call a dippy process! Having to resort to such hocus-pocus has prevented us from proving that the theory of quantum electrodynamics is mathematically self-consistent. It's surprising that the theory still hasn't been proved self-consistent one way or the other by now; I suspect that renormalization is not mathematically legitimate. What *is* certain is that we do not have a good mathematical way to describe the theory of quantum electrodynamics.

Value of Physical Reasoning

Feynman's account begins to wind down as he says[1]:

This completes the story of the development of the space-time view of quantum electrodynamics. I wonder if anything can be learned from it. I doubt it. It is most striking that most of the ideas developed in the course of this research were not ultimately used in the final result. For example, the half-advanced and half-retarded potential was not finally used, the action expression [for action at a distance] was not used, the idea that charges do not act on themselves was abandoned. The path integral formulation of quantum mechanics was useful for guessing at final expressions and at formulating the general theory of electrodynamics in new ways—although, strictly, it was not absolutely necessary. The same goes for the idea of the positron being a backward moving electron; it was very convenient, but not strictly necessary ... .

We are struck by the very large number of different physical viewpoints and widely different mathematical formulations that are all equivalent to one another. The method used here, of reasoning in physical terms, therefore, appears to be extremely inefficient. On looking back over the work, I can only feel a kind of regret for the enormous amount of physical reasoning and mathematical re-expression which ends by merely re-expressing what was previously known, although in a form which is much more efficient for the calculation of specific problems. Would it not have been much easier to simply work entirely in the mathematical framework to elaborate a more efficient expression? This would certainly seem to be the case, but it must be remarked that although the problem actually solved was only such a reformulation, the problem originally tackled was the (possibly still unsolved) problem of avoidance of the infinities of the usual theory. Therefore, a new theory was sought, not just a modification of the old. Although the quest was unsuccessful, we should look at the question of the value of physical ideas in developing a *new* theory ... .

I think the problem is not to find the *best* or most efficient method to proceed to a discovery, but to find any method at all. Physical reasoning does help some people to generate suggestions as to how the unknown may be related to the known. Theories of the known, which are described by different physical ideas may be equivalent in all their predictions and are hence scientifically indistinguishable. However, they are not psychologically identical when trying to move from that base into the unknown. For different views suggest different kinds of modifications which might be made .... . I, therefore, think that a good theoretical physicist today might find it useful to have a wide range of physical viewpoints and mathematical expressions of the same theory ... available to him. This may be asking too much of one man. Then new students should as a class have this. If every individual student follows the same current fashion in expressing and thinking about [the generally understood areas], then the variety of hypotheses being generated to understand [the still open problems] is limited. Perhaps rightly so, for possibly the chance is high that the truth lies in the fashionable direction. But [if] it is in another direction ... who will find it?

So spoke an honest man, the outstanding intuitionist of our age and a prime example of what may lie in store for anyone who dares to follow the beat of a different drum.

References

1. R. P. Feynman, "The Development of the Space-Time View of Quantum Electrodynamics," Nobel lecture, 11 December 1965, in *Les Prix Nobel en 1965*, Nobel Foundation, Stockholm (1966); edited version printed in PHYSICS TODAY, August 1966, p. 31.
2. R. P. Feynman, *"Surely You're Joking, Mr. Feynman!" Adventures of a Curious Character*, Norton, New York (1985).
3. R. P. Feynman, *QED*, Princeton U.P., Princeton, N. J. (1985).

This article was originally published in the February 1989 issue of *Physics Today*.

The
Research Physicist
at Cal Tech

DICK FEYNMAN—THE GUY IN THE OFFICE DOWN THE HALL

Murray Gell-Mann

I hope someday to write a lengthy piece about Richard Feynman as I knew him (for nearly 40 years, 33 of them as his colleague at Caltech), about our conversations on the fundamental laws of physics, and about the significance of the part of his work that bears on those laws. In this brief note, I restrict myself to a few remarks and I hardly touch on the content of our conversations.

When I think of Richard, I often recall a chilly afternoon in Altadena shortly before his marriage to the charming Gweneth. My late wife, Margaret, and I had returned in September 1960 from a year in Paris, London and East Africa; Richard had greeted me with the news that he was "catching up with me"—he too was to have an English wife and a small brown dog. The wedding soon took place, and it was a delightful occasion. We also met the dog (called Venus, I believe) and found that Richard was going overboard teaching her tricks (leading his mother, Lucille, with her dry wit, to wonder aloud what would become of a

Murray Gell-Mann is the Robert A. Millikan Professor of Theoretical Physics at the California Institute of Technology.

child if one came along). The Feynmans and we both bought houses in Altadena, and on the afternoon in question Margaret and I were visiting their place.

Richard started to make a fire, crumpling up pages of a newspaper and tossing them into the fireplace for kindling. Anyone else would have done the same, but the way he made a game out of it and the enthusiasm that he poured into that game were special and magical. Meanwhile, he had the dog racing around the house, up and down the stairs, and he was calling happily to Gweneth. He was a picture of energy, vitality and playfulness. That was Richard at his best.

He often worked on theoretical physics in the same way, with zest and humor. When we were together discussing physics, we would exchange ideas and silly jokes in between bouts of mathematical calculation—we struck sparks off each other, and it was exhilarating.

What I always liked about Richard's style was the lack of pomposity in his presentation. I was tired of theorists who dressed up their work in fancy mathematical language or invented pretentious frameworks for their sometimes rather modest contributions. Richard's ideas, often powerful, ingenious and original, were presented in a straightforward manner that I found refreshing.

I was less impressed with another well-known aspect of Richard's style. He surrounded himself with a cloud of myth, and he spent a great deal of time and energy generating anecdotes about himself.

Sometimes it did not require a great deal of effort. For example, during my first decade at Caltech there was a rule at our faculty club, the Athenaeum, that men had to wear jackets and ties at lunch. Richard usually came to work quite conventionally dressed (for those days) and hung his jacket and tie in his office. He rarely ate lunch at the Athenaeum, but when he did, he would often make a point of walking over in his shirt sleeves, tieless, and then putting on one of the ragged sport coats and one of the loud ties that the Athenaeum provided in the cloakroom for men who arrived unsuitably attired.

Many of the anecdotes arose, of course, through the stories Richard told, of which he was generally the hero, and in which he had to come out, if possible, looking smarter than anyone else. I must confess that as the years went by I became uncomfortable with the feeling of being a rival whom he wanted to surpass; and I found working with him less congenial because he seemed to be thinking more in terms of "you" and "me" than "us." Probably it was difficult for him to get used to collaborating with someone who was not just a foil for his own ideas (especially someone like me, since I thought of Richard as a splendid person to bounce my ideas off!).

At first, none of that was much of a problem, and we had many fine discussions in those days. In the course of those talks not only did we "twist the tail of the cosmos," but we also exchanged a good many lively reminiscences about our experiences in research.

Summing Over Histories

He told me, of course, of his graduate student days at Princeton and his adventures with his adviser, John Wheeler. Wheeler judged their work on the "absorber theory of radiation" to be too much of a collaboration to qualify as a dissertation for the PhD, and so Richard pursued his interest in Paul Dirac's work on the role of the action S in quantum mechanics. In his book on quantum mechanics, and even more in his article in the *Physikalische Zeitschrift der Sowjetunion* in 1932, Dirac had carried the idea quite far. He had effectively shown how a quantum mechanical amplitude for the transition from a set of values of the coordinates at one time to another set of values at a later time could be represented as a multiple integral, over the values of the coordinates at closely spaced intermediate times, of $\exp(iS/\hbar)$, where S is the value of the classical action along each sequence of intermediate coordinate values. What Dirac had not done was to state the result in so many words, to point out that this method could be used as the starting point for all quantum mechanics, and to suggest it as a practical way of doing quantum mechanical calculations.

Richard did just those things, I understand, in his 1942 dissertation, and then used the "path integral" or "sum over paths" approach in a great deal of his subsequent research. It was the basis, for example, of his way of arriving at the now standard covariant method of calculation in quantum field theory (which Ernst Stueckelberg reached in a different manner). That method is, of course, always presented in terms of "Feynman diagrams" such as the ones Dick later had painted on his van.

The sum-over-paths formulation is particularly convenient for integrating out one set of coordinates to concentrate on the remaining set. Thus the photon propagator in quantum electrodynamics is obtained[1] by "integrating out" the photon variables, leaving electrons and positrons, both real and virtual, to interact by means of the covariant function $\delta(x^2) + (\pi i x^2)^{-1}$.

In 1963 Feynman and his former student F. L. Vernon Jr, carrying further some research Ugo Fano had earlier done in a different way, showed how in a wide variety of problems of concern to laser physicists, condensed matter physicists and others of a practical bent, one can integrate out variables that are not of interest to throw light on the behavior of the ones that are kept. If initially the density matrix factors into one part depending on the interesting variables and another part depending on the rest, then the subsequent time development of the reduced density matrix for the interesting variables can be expressed in terms of a double path integral in which the coefficient of the initial reduced density matrix is $\exp[i(S-S'+W)/\hbar]$, where S is the action along the path referring to the left-hand side of the density matrix, S' is the action along the path referring to the right-hand side of density matrix, and W is the "influence functional," depending on both paths, that comes from integrating out all the uninteresting variables. Feynman and Vernon worked out a number of cases in detail, and subsequent research by A. O. Caldeira and Anthony Leggett, among others, further clarified some of the issues involved.

Shedding Light on Quantum Mechanics

More recently, in the work of H. Dieter Zeh, of Erich Joos and of Wojciech Żurek and others, this line of research has thrown important light on how quantum mechanics produces decoherence, one of the conditions for the nearly classical behavior of familiar objects. For a planet, or even a dust grain, undergoing collisions with, for example, the photons of the 3K radiation, the imaginary part of the functional W resulting from the integration over those quanta can yield, in $\exp(iW/\hbar)$, a factor that decreases exponentially with some measure of the separation between the coordinate trajectory on the left side of the density matrix and that on the right. The density matrix can thus be constrained to remain nearly diagonal in the coordinates of the particle, giving rise to decoherence. If in addition the dust grain's inertia is large enough that the grain resists, for the most part, the disturbances of its trajectory caused by the quantum and thermal fluctuations of the background, and also large enough that the quantum spreading of the coordinate is slow, then the behavior of the grain's position operator will be nearly classical.

When an operator comes into correspondence with a nearly classical operator, then the first operator can be measured or observed.

Thus work such as that of Feynman and Vernon has led not only to practical applications but also to a better understanding of how quantum mechanics produces the world with which we are familiar.

The path integral approach has proved in numerous situations to be a useful alternative to the conventional formulation of quantum mechanics in terms of operators in Hilbert space. It has many advantages besides the ease of integrating out, under suitable conditions, some of the variables. The path integral method, making use as it does of the action, can usually display in an elegant manner the invariances of the theory and can point the way toward exhibiting those invariances in a perturbation expansion. It is obviously a good approach for deriving the classical limit, and it can also be very helpful in semiclassical approximations, for example, in the description of tunneling. For certain effects, such as tunneling via instantons, it permits calculations that are highly nonperturbative in the usual sense. It is also particularly good for the global study of field configurations in quantum field theory, as it permits a straightforward discussion of topological effects.

Of course the conventional approach is superior for certain purposes, such as exhibiting the unitarity of the S matrix and the fact that probabilities are not negative. Richard would never have contemplated, as he did around 1948, the consistent omission of all closed loops in quantum electrodynamics if he had been thinking in terms of a Hamiltonian formulation, where unitarity, which rules out such an omission, is automatic. (The impossible theory without closed loops could, by the way, realize the remarkable vision of Wheeler, which Richard said Wheeler once awakened him to explain: Not only are positrons electrons going backward in time, but all electrons and positrons represent the same electron going backward and forward, thus explaining why they all have the same absolute value of the electric charge!)

In any case, the path integral formulation remained merely a reformulation of quantum mechanics, equivalent to the usual formulation. I say "merely" because Richard, with his great talent for working out, sometimes in dramatically new ways, the consequences of known laws, was unnecessarily sensitive on the subject of discovering new ones. He wrote, in connection with the discovery of the universal vector and axial vector weak interaction in 1957: "It was the first time, and the only time, in my career that I knew a law of nature that nobody else knew. (Of course, it wasn't true, but finding out later that at least Murray Gell-Mann—and also [E. C. George] Sudarshan and [Robert] Marshak—had worked out the same theory didn't spoil my fun.) ... It's the only time I ever discovered a new law."[2]

Thus it would have pleased Richard to know (and perhaps he did know, without my being aware of it) that there are now some indications that his PhD dissertation may have involved a really basic advance in physical theory and not just a formal development. The path integral formulation of quantum mechanics may be more fundamental than the conventional one, in that there is a crucial domain where it may apply and the conventional formulation may fail. That domain is quantum cosmology.

Seeking Rules for Quantum Gravity

Of all the fields in fundamental physical theory, the gravitational field is picked out as controlling, in Einsteinian fashion, the structure of space-time. This is true even in a unified description of all the fields and all the particles of nature. Today, in superstring theory, we have the first respectable candidate for such a theory, apparently finite in perturbation theory and describing, roughly speaking, an infinite set of local fields, one of which is the gravitational field linked to the metric of space-time. If all the other fields are dropped, the theory becomes an Einsteinian theory of gravitation.

Now the failure of the conventional formulation of quantum mechanics, if it occurs, is connected with the quantum mechanical smearing of space-time that is inevitable in any quantum field theory that includes Einsteinian gravitation.

If there is a dominant background metric for space-time, especially a Minkowskian metric, and one is treating the behavior of small quantum fluctuations about the background (for example, the scattering of gravitons by gravitons), then the deep questions about space-time in quantum mechanics do not come to the fore.

Dick played a major part in working out the rules of quantum gravity in that approximation. It so happened that I was peripherally involved in the story of that research. We first discussed it when I visited Caltech during the Christmas vacation of 1954–55 and he was my host. (I was offered a job within a few days—such things would take longer now.) I had been interested in a similar approach, sidestepping the difficult cosmological issues, and when I found that he had made considerable progress I encouraged him to continue, to calculate one-loop effects and to find out whether quantum gravity was really a divergent theory to that order. He was always very suspicious of unrenormalizability as a criterion for rejecting theories, but he did pursue the research on and off. In 1960 he complained to me that he was having trouble. His covariant diagram method was giving results incompatible with unitarity. The imaginary part of the amplitude for a fourth-order process should be related directly to the product of a second-order amplitude and the complex conjugate of a second-order amplitude. That relation was failing.

I suggested that he try the analogous problem in Yang-Mills theory, a much simpler nonlinear gauge theory than Einsteinian gravitation. Richard asked what Yang-Mills theory was. (He must have forgotten, because in 1957 we worked out the coupling of the photon to the charged intermediate boson for the weak interaction and noticed that it was the right coupling for a Yang-Mills theory of those quanta.) Anyway, it didn't take long to teach him the rudiments of Yang-Mills theory, and he threw himself with renewed energy into resolving the contradiction. He found, eventually, that in the Lorentz-covariant formulation of either theory it was necessary to introduce some weird supplementary fields called "ghosts," and they have been used ever since, acquiring more and more importance. He described them at a meeting in Poland (in 1963, I think). Usually they are called "Faddeev-Popov ghosts" after L. D. Faddeev and V. N. Popov, who also studied them.

Thus Feynman was able to report in the 1960s that Einsteinian gravitation was terribly divergent when interacting with electrons, photons or other particles. (The divergences in *pure* quantum gravitation theory turned out to be serious too, but that was shown much later, in the two-loop approximation, by two Caltech graduate students, Marc Goroff and Augusto Sagnotti.)

Those problems may be rectified by unification of all the particles and interactions, as they are in superstring theory. But we must still face up to the issues raised by the fact that the metric is up for quantum mechanical grabs and cannot in general be treated as a simple classical background plus small quantum fluctuations.

Quantum Cosmology

Recently there has been great progress in thinking about the cosmological aspects of quantized Einsteinian gravitation. The work of Stephen Hawking and James Hartle, as well as Claudio Teitelboim, Alexander Vilenkin, Jonathan Halliwell and several others, has shown how the path integral method can probably deal with the situation and how it may be possible to generalize the method so as to describe *not only the dynamics of the universe but also its initial boundary condition* in terms of the classical action S. Furthermore, there are now, as I mentioned above, some indications that the conventional formulation of quantum mechanics may not be justifiable except to the extent that a background space-time emerges with small quantum fluctuations. Hartle in particular has emphasized such a possibility.

One crude way to see the argument is to express the wavefunction of the universe (which we assume to be in a pure state) as a path integral over all the fields in nature (for example, the infinity of local fields represented, roughly speaking, by the superstring), reserving the integral over the metric $g_{\mu\nu}$ for last. The total action S can be represented as the Einstein action S_G for pure gravitation plus the actions S_M for all the other, "matter" fields, including their coupling to gravitation. We have, then, crudely,

$$\text{Amplitude} = \int \mathscr{D}g_{\mu\nu} \exp \frac{iS_G}{\hbar}$$

$$\times \int \mathscr{D}(\text{everything else}) \exp \frac{iS_M}{\hbar}$$

For the moment, suppose only $g_{\mu\nu}$ configurations corresponding to a simple topology for space-time are allowed.

Before the integration over $g_{\mu\nu}$ is performed, there is a definite space-time, with the possibility of constructing well-defined space-like surfaces in a definite succession described by a time-like variable. There is an equivalent Hilbert-space formalism; we have unitarity (conservation of positive probability); and we can have conventional causality (it corresponds in the Hilbert-space formulation to the requirement of time ordering of operators in the formula for probabilities).

Now, when the integral over $g_{\mu\nu}$ is done, it is no longer clear that any of that machinery remains, since we are integrating over the structure of space-time and once the integral is performed it is hard to point to space-like surfaces or a succession described by a time-like variable. Of course it may be possible to construct a Hilbert-space formulation, with unitarity and causality, in some new way, perhaps employing a new, external time variable of some kind (what Feynman liked to call a fifth wheel), but it is by no means certain that such a program can be carried out.

At this stage, we may admit the possibility of summing over all topologies of space-time (or of the corresponding space-time with a Euclidean metric). If that is the correct thing to do, then we are immediately transported into the realm of baby universes and worm-holes, so beloved of Stephen Hawking and now so fashionable, in which it seems to be demonstrable that the cosmological constant vanishes. In that realm the path integral method appears able to cope, and it remains to be seen to what extent the conventional formulation of quantum mechanics can keep up.

For Richard's sake (and Dirac's too), I would rather like it to turn out that the path integral method is the real foundation of quantum mechanics and thus of physical theory. This is true despite the fact that, having an algebraic turn of mind, I have always personally preferred the operator approach, and despite the added difficulty, in the absence of a Hilbert-space formalism, of interpreting the wavefunction or density matrix of the universe (already a bit difficult to explain in any case, as anyone attending my classes will attest). If notions of transformation theory, unitarity and causality really emerge from the mist only after a fairly clear background metric appears (that metric itself being the result of a quantum mechanical probabilistic process), then we may have a little more explaining to do. Here Dick Feynman's talents and clarity of thought would have been a help.

Turning Things Around

Richard, as is well known, liked to look at each problem, important or unimportant, in a new way—"turning it around," as he would say. He told how his father, who died when he was young, taught him to do that. This approach went along with Richard's extraordinary efforts to be different, especially from his friends and colleagues.

Of course any of us engaged in creative work, and in fact anyone having a creative idea even in everyday life, has to shake up the usual patterns in some way in order to get out of the rut (or the basin of attraction!) of conventional thinking, dispense with certain accepted but wrong notions, and find a new and better way to formulate some problem. But with Dick, "turning things around" and being different became a passion.

The result was that on certain occasions, in scientific work or in ordinary living, when an imaginative new way of looking at things was needed, he could come up with a remarkably useful innovation. But on many other occasions, when the usual way of doing business had its virtues, he was not the ideal person to consult. Remember his television appearance in which he made fun of the daily habit of brushing one's teeth? (And he didn't even suggest flossing!) Or take his occasional excursions into far-out political choices in the 1950s, during his second marriage. Those certainly set him off from most of his friends. But one day during that time, he called me and sheepishly admitted having voted for a particularly outrageous candidate for statewide office—and then asked me if in the future I would check over such names beforehand and tell him when he was really going off the deep end!

None of the aberrations mentioned here changes the fact that Dick Feynman was a most inspiring person. I have referred to his originality and straightforwardness and to his energy, playfulness and vitality. All of those characteristics showed up in his work and also in the other facets of his life. Indeed, that vitality may be related to the kind of biological (and probably psychological) vitality that enabled him to resist so remarkably and for so long the illness to which he finally succumbed.

When I think of him now, it is usually as he was during that first decade that we were colleagues, when we were both young and everything seemed possible. We phoned each other with good ideas and crazy ones, with serious messages and farcical gags. We yelled at each other in front of the blackboard. We taught stewardesses to say "quark-quark scattering" and "quark-antiquark scattering." We delivered a peacock to the bedroom of our friend Jirayr Zorthian on his birthday, while our wives distracted him. We argued about everything under the sun.

Later on, we drifted apart to a considerable extent, but I was aware, all the time we were colleagues, that if a really profound question in science came up, there would be fun and profit in discussing it with Dick. Even though on many occasions during the last 20 years, I passed up the opportunity to talk with him in such a case, I knew that I *could* do so, and that made a great difference.

Besides, I did not always pass it up. For example, during the last few months and even weeks of his life, we kept up a running discussion of one of the most basic subjects, the role of "classical objects" in the interpretation of quantum mechanics. We thus resumed a series of conversations on that topic that we had begun a quarter of a century earlier. In between 1963–64 and 1987 those talks about quantum mechanics were rare, but there was at least one remarkable occasion during the last few of those years. Richard sat in on one of my classes on the meaning of quantum mechanics, interrupting from time to time. He did not, however, object to what I was saying; rather, he reinforced the points I was making. The students must have been delighted as they heard the same arguments made by both of us in a kind of counterpoint.

It is hard for me to get used to the fact that now, when I have a deep issue in physics to discuss with someone, Dick Feynman is no longer around.

I should like to thank James B. Hartle for many instructive conversations about quantum mechanics and the path integral method in quantum cosmology.

References

1. R. P. Feynman, Phys. Rev. **76**, 749, 769 (1949).
2. R. P. Feynman, *"Surely You're Joking, Mr. Feynman!" Adventures of a Curious Character*, Bantam, New York (1986), p. 229. See also M. Gell-Mann, in *Proc. Int. Mtg. on the History of Scientific Ideas*, M. G. Doncel *et al.*, eds., Bellaterra, Barcelona (1987), p. 474.

This article was originally published in the February 1989 issue of *Physics Today.*

FEYNMAN AND PARTONS

James D. Bjorken

For me, as for so many others, Richard Feynman is a special hero. He became so while I was learning quantum electrodynamics in graduate school at Stanford. The course happened to be organized historically, and for several months it was taught in the 1930s style out of Heitler's classic text, using old-fashioned perturbation theory and Dirac matrices α and β (but not γ). After this trial by fire came a seemingly endless, gloomy, turgid mass of field-quantization formalism. When Feynman diagrams arrived, it was the sun breaking through the clouds, complete with rainbow and pot of gold. Brilliant! Physical and profound! It was instant conversion to discipleship.

For many years thereafter my discipleship developed like most everyone else's, through the strong influence of his writings and the occasional rare treat of hearing him perform live. But it became my privilege that for a few years my research path ran in parallel with his. This convergence came about because of the remarkable and historic series of inelastic electron scattering experiments at the Stanford Linear Accelerator Center by a SLAC-MIT collaboration. These experiments played a crucial role in revealing the existence of point-like, quark constituents of the proton, while Feynman's insights and intu-

James D. Bjorken is a permanent staff member, Stanford Linear Accelerator Center.

ition provided much of the theoretical motive power for the interpretation of the experimental developments.

In the late 1960s, when the SLAC program was initiated, Feynman was working on descriptions of high-energy hadron-hadron collisions. He pictured the typical reaction as occurring by the exchange of constituents—Feynman called them partons—between the rapidly moving projectiles. The primary basis for his parton picture was empirical; significant evidence was the apparently exponentially bounded transverse-momentum distribution of produced or scattered secondary particles. This indicated a predominantly "soft" interaction; that is, the important dynamics occurred at an intrinsic distance scale on the order of the proton size. Exchange of constituents satisfied this "softness" criterion very well. Indeed there was no explicit interaction introduced at all, only the implicit one constraining the constituents to be within the proton.

Inclusive Processes

In those days, using local field theory to describe the strong force was no more fashionable than using it nowadays to describe quantum gravity. Rather, those were the glory days for Regge-pole theorists. It was believed that the processes important for detailed study were ones with no more than two particles in the final state. The high-energy limit of the cross section for such collisions is the natural domain of applicability of the Regge-pole theory, which need not be elaborated here in detail. Feynman's partons provided a novel way to interpret the Regge-pole picture. But more important was Feynman's introduction of a new language for describing inelastic collisions involving the production of many, not just two, particles.

Multiparticle collisions were in those days largely shunned by theorists, who preferred to study only processes in which all the final particles are observed and all the momenta determined. Feynman called such processes "exclusive," and he emphasized, by contrast, "inclusive" processes, in which one (or a few) particles in the final state are identified and their momenta specified, but all other possibilities are summed over. Such processes were largely unknown to theorists, although the experimental community knew measurements of inclusive distributions as "beam surveys": chores required when commissioning a new accelerator to ensure proper design and implementation of secondary beam lines and radiation shielding. Feynman suggested that the inclusive distributions were themselves worthy of theoretical attention and suggested a scaling behavior in the variable x_F, the ratio of longitudinal momentum of a secondary particle to the maximum value allowed by energy-momentum conservation. He also emphasized rapidity (essentially the logarithm of the particle momentum) as an especially useful variable and argued that the distribution of particles produced in high-energy collisions was essentially uniform in that variable.

But this initial motivation for Feynman's partons was soon replaced by a stronger one. It happened almost by chance. Feynman was visiting his sister in the San Francisco Bay area and happened to stop by SLAC for a short visit. He was shown the latest electron-proton scattering data, along with fits to a scaling law I had suggested to the experimentalists. I was out of town, and a puzzled Feynman did not get a clear picture from the experimentalists of where the scaling law originated: "something about current algebra, sum-rules, Regge-theory ... ."

It took Feynman only an evening of calculation with his partons to interpret what was going on. He viewed the process in a reference frame in which the motion of the target proton was extremely relativistic. In that frame the proton was replaced, as in his previous calculations, by a "beam" of its constituents, or partons. He assumed the electron scattered elastically and incoherently from these partons, which he regarded as point-like quanta with no interactions among them. Feynman viewed the scaling function I had introduced as giving the probability of finding a parton of a given momentum in the *incident* proton beam, weighted by the square of the parton electric charge.

As I recall, I returned to SLAC just before Feynman was to leave and found much excitement within—and beyond—the theory group there. Feynman sought me out and bombarded me with queries. *"Of course you must know this ... . Of course you must know that ...,"* he kept saying. I knew about some of the things Feynman mentioned; others I didn't know. And there were things that I knew at the time but he did not. What I vividly remember was the language he used: It was not unfamiliar, but it was distinctly *different*. It was an easy, seductive language that everyone could understand. It took no time at all for the parton model bandwagon to get rolling.

Feynman's calculation of the electron-proton scattering cross section invited generalization to many electromagnetic and weak processes. Feynman continued to develop the ideas at Caltech; I worked with Emmanuel Paschos and others at SLAC; and a horde of others joined in. Ideas and methods were developed for determining the parton spin, charge and weak-interaction properties, and with time the natural identification of (charged) partons with quarks became established. Central to this course of events were the experiments, especially the elegant, sophisticated series of electron-proton scattering measurements by the SLAC-MIT collaboration,[1] as well as the data from neutrino experiments at CERN and Fermilab.

As the quark-parton model took hold, an immediate problem arose: Why (or, at the very least, *how*) no fractional charge was seen in the collision debris. These questions became a major topic for Feynman and for me. And it was in this "second generation" evolution of the parton model that my scientific life ran most in parallel with Feynman's. It became an ongoing challenge for me to figure out how he was thinking about a given problem or how he would think about that problem if he got around to it. On occasion this attitude helped in finding the solution. Only rarely did we directly communicate and compare notes—although I did sometimes get indirect information from others who had made the pilgrimage to Caltech.

In one instance I had a tangible measure of success in my attempt to follow Feynman's ways. In a review talk on partons and related issues, I cited "Feynman's notebooks" at a rate of about one citation per transparency, for I suspected that he had worked out all kinds of things but not published his results. The audience loved it. But not only did I not really know what Feynman knew and when he knew it, I did not even realize that he *kept* notebooks. (I am told that there exist very careful and complete logbooks, cross-referenced, of his day-by-day work). Sometime later I had the opportunity to reminisce with him about that talk, and he confirmed that with one exception (I forget what it was), it was all there. That unpublished work included light-cone quantization (with some sophisticated applications to QED), independent work on operator-product expansions, and his "fluid analogy," which compared the properties of parton, as well as produced-hadron, distributions in relativistic phase space with those of ordinary fluids (having short-range correlations only) in configuration space. (The fluid-analogy ideas were revealed to the outside world by Kenneth Wilson.)

Deductive vs Inductive Thinking

During this parallel interaction with Feynman, there occurred a strong influence on my style of thinking in physics. The problems that the parton model raised were not to be solved using the methods one learns in Physics 101. Characteristically, Feynman addressed the fundamental issues raised by the parton model very directly right from the start. For example, he wrote in his first paper on partons:[2]

> These suggestions arose in theoretical studies from several directions and do not represent the result of consideration of any one model. They are an extraction of those features which relativity and quantum mechanics and some empirical facts imply almost independently of a model. I have difficulty in writing this note because it is not in the nature of a deductive paper, but is the result of an induction. I am more sure of the conclusions than of any single argument which suggested them to me for they have an internal consistency which surprises me and exceeds the consistency of my deductive arguments which hinted at their existence.

The power of the parton model came not from a linear, deductive logical line such as one finds in an ordinary computer, but rather from a multidimensional logical network more typical of the human brain. And this situation applied not only to the creative process, where it is not uncommon, but also to the end product. It was the inner consistency of a broad variety of lines of attack that was impressive. One may legitimately question this house-of-cards approach to science: One good argument is better, after all, than 52 mutually supporting inferior ones.

I came to realize, in fact, that in my work leading to the ideas of scaling in inelastic electron scattering this inductive approach had also predominated. But I was a young postdoc at that time, and I had little confidence in it: to me a claimed result required a clean line of logic (even were it to be constructed *ex post facto*) in order to meet the standards of the trade. And such logical lines were hard to find.

The situation was clearly present for Feynman as well. His original journal article[2] on parton-model ideology nowhere mentions the word parton or proton constituent. The parton was introduced only in a less formal conference talk given at about the same time.[3] And even years later, in his book[4] *Photon-Hadron Interactions*, the ambivalence still appears. The concluding pages of that book contain the following phrases:

We have built a very tall house of cards making so many weakly-based conjectures one upon the other and a great deal may be wrong ... .

Finally it should be noted that even if our house of cards survives and proves to be right we have not thereby proved the existence of partons ... .

From this point of view the partons would appear as an unnecessary scaffolding that was used in building our house of cards.

On the other hand, the partons would have been a useful psychological guide as to what relations to expect—and if they continued to serve this way to produce other valid expectations they would of course begin to become "real," possibly as real as any other theoretical structure invented to describe nature.

Of these phrases, the last one has turned out to be the most prophetic.

It is hard to document here the reasons for trusting the parton ideology. Many of the results seemed to be based on broad principles of a mostly kinematical nature. For example, a main feature of the parton picture is the remarkably nonrelativistic character of the extreme relativistic limit. Not only do the internal motions of constituents of a high-momentum hadron slow down because of relativistic time dilation, but the transverse dynamics really does look nonrelativistic. This can already be glimpsed from the energy-momentum relation for a free particle moving rapidly in the z direction,

$$E^2 = c^2(p_z^2 + p_x^2 + p_y^2) + m^2 c^4$$

rewritten as the Hamiltonian for the transverse dynamics

$$\mathsf{H} = E - p_z c = \frac{p_x^2 + p_y^2}{2\eta} + \frac{m^2 c^4}{2\eta} ,$$

where

$$\eta = \frac{E + p_z c}{2c^2}$$

represents *inertia*, in proportion (as $E, p_z \to \infty$) to the total momentum of the particle. This analogy invited a qualitative, intuitive view of the problem, abstracted from nonrelativistic quantum theory.

But this nonrelativistic intuition about dynamics in the extreme-relativistic limit was only one line of attack. Another was the consistency, sometimes hard won, of the proposed answers when the dynamics of the processes were studied in a variety of reference frames. Yet another was the smooth matching of the predictions for, say, a given inclusive process with the expectations for the set of exclusive processes comprising that inclusive process. The buzz-word for this criterion is duality.

The net result of such second-generation attempts to understand the final states in these hard-collision parton processes turned out to be remarkably unremarkable: These processes should look essentially the *same* as ordinary collisions at the same available center-of-mass energy. This was *ab initio* not obvious to the theoretical community. Because of the unspectacular nature of this result, the data supporting it created little stir in the experimental community. But for me, and I suspect for Feynman also, the experimental results were deeply satisfying.

It is worth emphasizing again that throughout this development of the parton model the essential input assumption about the dynamics was that strong interactions were "soft," that is, characterized by a force whose range was about the same as the proton size. As it turns out, this assumption is not quite right. The currently accepted theory of the strong force, quantum chromodynamics, contains, in addition to the strong, soft interaction, a not-so-strong hard interaction that becomes significant at much shorter distance scales. The latter is analogous to the inverse-square electromagnetic force but with a fine-structure constant of about 1/7. Long before QCD emerged on the scene, the possibility of such a hard strong interaction was entertained. Feynman was always careful to set this hypothesis separate from those of the basic parton model. As the evidence for QCD grew, Feynman (with Richard Field) worked out the modifications to the "naive" parton model phenomenology implied by QCD, and grappled with the fundamental properties of QCD that might explain confinement. By now the basic parton model concepts have been deeply integrated into the formalism of QCD, to the extent that most theorists take the parton picture to be a self-evident consequence of QCD. I suspect there is more to the story than that, yet to be uncovered. But it consists of questions of rigor and of detail; the parton approach will not become obsolete.

With the emergence of QCD, my interests drifted apart from Feynman's. Even during the period when we had common interests, I had relatively little personal contact with Feynman. Our relationship was warm, but it was not closely personal. It is not that I don't feel close to Feynman. Something of him is very much in me and always will be. And I will always treasure that.

References

1. For a general review, see J. I. Friedman, H. W. Kendall, Ann. Rev. Nucl. Sci. **22**, 203 (1972).
2. R. P. Feynman, Phys. Rev. Lett. **23**, 1415 (1969).
3. R. P. Feynman, in *Proc. III Int. Conf. on High-Energy Collisions*, organized by C. N. Yang *et al.*, Gordon and Breach, New York (1969).
4. R. P. Feynman, *Photon-Hadron Interactions*, Benjamin, Reading, Mass. (1972).

This article was originally published in the February 1989 issue of *Physics Today*.

RICHARD FEYNMAN AND CONDENSED MATTER PHYSICS

David Pines

From 1953 to 1958 Richard Feynman worked primarily on problems in condensed matter physics. Of the 14 scientific papers he published during this period, ten are devoted to the physics of liquid helium, one discusses the relation between superconductivity and superfluidity, and one deals with the motion of slow electrons in polar crystals, the "polaron" problem; the remaining two describe work Feynman had carried out on quantum electrodynamics and hadron physics earlier at Cornell. He brought to the condensed matter problems the same remarkable originality and physical insight that characterized his earlier work on quantum electrodynamics and the path integral method, and through his contributions he made a lasting impact on the subfields of low-temperature physics and statistical mechanics.

Here I shall focus primarily on his contributions to the theory of liquid helium. Because they offer good insight into how Feynman approached a new problem, I shall also discuss briefly his ideas about superconductivity (which for Richard was "the one that got away"), his work on polarons and his lectures on statistical mechanics.

David Pines is Center for Advanced Study Professor of Physics at the University of Illinois, Urbana-Champaign.

About 35 years ago, the condensed matter and low-temperature community of theorists was a small one; we all knew one another personally, listened to one another's lectures and discussed problems together whenever we met. Where possible, I shall try to convey some sense of what it was like to have had Richard Feynman as a member of that community.

However, just as it is not possible to capture in words the experience of listening to Feynman discuss his research, it also is not possible to capture the flavor of his written work. Reading Feynman is a joy and a delight, for in his papers, as in his talks, Feynman communicated very directly, as though the reader were watching him derive the results at the blackboard. Thus, for those of us who had the pleasure of knowing Feynman, his papers bring vividly to life those discussions and lectures. He is explicit about how he has formulated the problem and what methods he has tried; he makes no attempt to gloss over difficulties; and he takes the reader fully into his confidence on such matters as research strategy, physical pictures vs mathematical calculations, unsolved aspects of a problem, promising approaches to their solution and so on. I cannot encourage the readers of this article too strongly to go forth and read the original; there is no substitute.

Liquid Helium Theory Prior to 1953

To put Feynman's contributions to our understanding of liquid helium into perspective, it is useful to recall the work of Fritz London, Lazlo Tisza and Lev Landau. London[1] proposed that the λ transition between normal liquid helium, He I, and the superfluid liquid, He II, at 2.19 K had its physical origin in the formation of a Bose condensate, a state in which all the atoms in the liquid would be in a single quantum state at absolute zero. The mechanism was analogous to the condensation of an ideal Bose-Einstein gas. Tisza[2] proposed a phenomenological two-fluid model to explain the behavior of He II, consisting of a superfluid component of density ρ_s that flowed without resistance and was the sole component at absolute zero, and a normal component of density ρ_n that resembled an ordinary liquid or even a gas. The proportion of normal fluid increased with increasing temperature until, at the λ point, ρ_s was 0 and ρ_n was ρ, the helium density. Landau[3] arrived at a similar point of view by developing a theory of quantum hydrodynamics, in which the normal fluid corresponded to a gas of two kinds of interacting excitations: phonons, which at long wavelengths are the

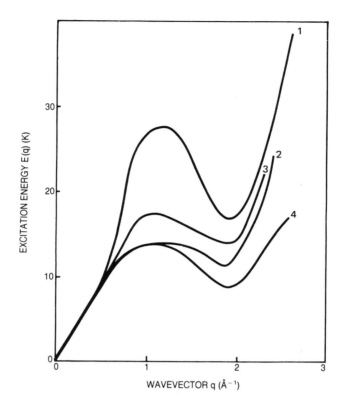

FIGURE 1 *Elementary excitation spectrum of liquid helium II as computed by Feynman[8] (1), Feynman and Cohen[12] (3), and the FC spectrum as corrected by Manousakis and Pandharipande[16] (2). The spectrum as measured by Woods and Cowley[14] is also shown (4).*

usual quantized sound waves of a compressible liquid, and rotons, short-wavelength excitations (corresponding to a momentum $p \sim p_0 \sim 2$ Å^{-1}) possessing a finite energy Δ (~ 10 K). Landau obtained the explicit form of the roton spectrum by analyzing early experiments on specific heat and second sound in He II; his proposed spectrum had the form shown in Figure 1.

Feynman remarked that London's idea "could be criticized on the grounds that the strong forces of interaction between the helium atoms might make the ideal-gas approximation even qualitatively incorrect,"[4] while "Tisza's view is frankly phenomenological"[5]; further, "the role of statistics in [Landau's] arguments is not clear,"[5] and, "since the rotons appear to correspond to only a few atoms, a complete understanding of the roton state can therefore only be achieved by way of an atomic viewpoint." Feynman then suggested that "a more complete study of liquid helium from first principles might attempt to answer at least three important questions: (a) Why does the liquid make a transition between two forms, He I and He II? (b) Why are there no states of very low energy, other than phonons which can be excited in helium II (*i.e.*, below 0.5 K)? (c) What is the nature of the excitations which constitute the 'normal fluid component' at higher temperatures, say from 1 to 2.2 [K]?" With these words Feynman described his research program on helium, a program he carried to completion during the period 1953–57.

An Atomic Theory of the λ Transition

To demonstrate that the strong interaction between helium atoms would not change the central features of the Bose condensation proposed by London, Feynman drew upon his space-time approach to quantum mechanics[6] to write the exact partition function as an integral over trajectories.[4] He used this expression to examine the character of the most important trajectories and concluded that despite the nearly hard-sphere character of the atomic interaction, the motion of a given atom would be little affected by the motions of others; the latter atoms would simply move out of its way, and so contribute to its effective mass, but would otherwise affect its motion little. Feynman was thus led to an approximate form of the partition function, which he analyzed in some detail to show that as the temperature is lowered, there must occur a transition that depends in an essential way on the Bose statistics of the helium atoms and that resembles closely the third-order transition (specific heat continuous, but possessing a discontinuous slope) found for the ideal Bose gas. He further noted that it would not be difficult to obtain a transition temperature of approximately 2 K and that the geometric correlations he neglected would likely turn the transition into a second-order one, in agreement with experiment.

Thirty-two years passed before Feynman's program was realized; in 1985 David Ceperley and Eugene Pollock[7] used a Cray-1 computer to carry out the necessary path integrals and obtained excellent agreement with experiment for the specific heat (and other properties) through and below the λ transition.

Viewed in perspective, this first paper by Feynman on liquid helium displays clearly that blend of magic, mathematical ingenuity and sophistication, and physical insight that is almost uniquely Feynman's. In reading this and his later papers one is struck by how differently Feynman's mind worked from that of other great physicists who studied—and in totally different ways solved—the same problem. Feynman took a very hands-on, direct computational point of view; Lars Onsager relied on field theory and Landau on extremely general arguments. Yet all came in the end to similar conclusions.

Feynman on Low-Lying Excited States

Feynman followed the work discussed above with a paper in which he examined the nature of the ground-state wavefunction of liquid He II and the character of the low-lying excitations. This paper is a *tour de force* in that it contains only a single equation (for the changes in free energy when a He^4 atom is replaced by He^3); rather it contains a series of closely reasoned arguments that led Feynman, and lead the reader, to the conclusion that because liquid He^4 obeys Bose statistics, there can be no low-lying excitations other than longitudinal phonons.

He began the paper with a qualitative description of the ground-state wavefunction, noting that if one freezes the motion of all the atoms at a given time, the wavefunction amplitude is negligible when any two atoms overlap (because of the strong hard-core-like repulsion) and is a maximum when a given atom is at the center of the "cage" formed by its neighbors. He then showed that extremely-low-energy excited states must involve large groups of atoms, and that compressional waves obeying a dispersion relation $\omega = sq$ (where ω is the frequency, s the sound velocity and q the wavevector) give a true mode of excitation. He demonstrated that this situation would not arise for an ideal Fermi gas or a classical (Boltzmann) gas, in both of which single-particle excitations yield the dominant low-lying mode. He clinched his case by showing that low-lying single-particle excitations cannot exist for a Bose liquid, because in this system the motion of a given atom from one location to another is "merely an interchange of which atom is which" and "cannot change the wavefunction." Feynman went on to examine the possible nature of "higher energy" excitations corresponding to the motion of a single atom or a small group of atoms, and argued that these might well correspond to Landau's rotons. He concluded the paper by introducing some of the themes to which he would return in his subsequent papers: the existence of a critical velocity for superfluid flow, and the motion of a small sphere or of a He^3 atom in the liquid.

A Wavefunction for Phonons and Rotons

In the third paper of the series,[8] Feynman extended the physical arguments of the second to show that the wavefunction that represents an excitation in He II must be of the form

$$\Psi_{exc} = \Psi_0 \sum_j f(r_j) , \tag{1a}$$

where Ψ_0 is the ground state wavefunction, f is some function of position, r, and the sum is over all the atoms. He then determined the form of $f(r)$ and the corresponding excited-state energy using the variational principle. He found $f(r) = \exp(i\mathbf{q} \cdot \mathbf{r})$, so that for an excitation of momentum $\mathbf{q}$ the wavefunction[9] is

$$\Psi_{exc} = \Psi_0 \sum_j \exp(i\mathbf{q} \cdot \mathbf{r}_j) \equiv \rho_q^+ \Psi_0 \tag{1b}$$

while the corresponding energy is

$$E_q = q^2/2mS_q \tag{2}$$

Here ρ_q^+ is the Fourier transform of the density, $\rho(r)$, while S_q, the Fourier transform of the two-particle correlation function, may be written as

$$S_q = \langle \Psi_0 | \rho_q^+ \rho_q | \Psi_0 \rangle / N, \tag{3}$$

where N is the number density of helium atoms. In this form it is seen that S_q is the liquid structure factor that determines the elastic scattering of neutrons or x rays and hence can be determined experimentally; for large q it approaches unity, while for small q, since ρ_q describes a phonon, it takes the form $S_q = q/2ms$, where s is the sound velocity. As shown in Figure 2, for values of $q \sim 2$ Å^{-1}, S_q reaches a maximum, so that the corresponding value of the excitation energy, shown in Figure 1, possesses a local minimum; the excitations in the vicinity of this minimum are Landau's rotons. Their energy Δ is, however, too large—some 18 K instead of the roughly 9.6 K that Landau found would fit the specific heat and second-sound velocity measure-

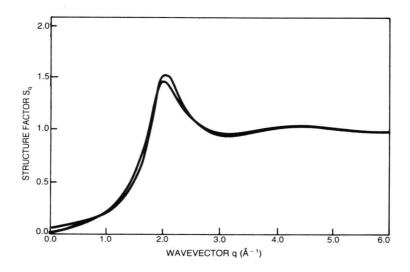

FIGURE 2. *Experimental results for the static structure factor for liquid helium. The black curve shows results for 2.29 K, the grey curve for 1.06 K.*

ments. Feynman found this result discouraging, and attributed it to the inaccuracy of the wavefunction (equation 1b) in this momentum region.

Feynman went on in this paper to consider the thermodynamic properties of He II based on his form for the excitation spectrum, noting (quite correctly) that as one goes to temperatures near the λ point, the number of excitations becomes sufficiently large that their interactions must be taken into account in any correct calculation. He then considered the motion of the fluid as a whole, noting that for irrotational motion the wavefunction must be of the form

$$\Psi = \Psi_0 \left\{ \exp\left[i \sum_j S(r_j) \right] \right\} , \qquad (4)$$

where $S(r)$ is some function of position, and the velocity of the fluid is given by

$$v_s(r) = \nabla S(\mathbf{r})/m \qquad (5)$$

so that the motion is indeed irrotational, that is, $\nabla \times v_s = 0$. He pointed out that one can have velocity fields that are not curl free, provided one considers regions that are not simply connected, as in a torus, because

S then need not be single valued; he further remarked that under these circumstances, states of angular momentum in multiples of $N\hbar$ become possible—a clear indication that he was beginning to think about vorticity.

On considering excitations in a moving fluid, Feynman obtained the familiar result $E=E_q+\mathbf{q}\cdot\mathbf{v}_s$, where $\mathbf{v}_s$ is the superfluid velocity, and used this to establish the relation between his excitations and the normal-fluid density, ρ_n. He obtained the Landau results, and shared with the reader his unease about what the mathematically correct separation of the current into two parts is, using a gas of interacting rotons moving in a background fluid to illustrate his concern about how this separation is to be accomplished.

In the process of examining excitations in the moving fluid, Feynman identified a further problem with his excited-state wavefunction: It does not lead to particle conservation when one forms a wave packet to describe an excitation that carries current $\hbar\mathbf{q}/m$ and drifts at a group velocity $\nabla_q E_q$ (the roton, for which the group velocity is vanishing, being an extreme example). He therefore proposed a qualitative improvement in his wavefunction, in which the return flow of the background liquid about a moving excitation acts to conserve particles. This *backflow*, he noted, should be dipolar at large distances from the excitation, and the coupling of, for instance, the roton to this return field should both lead to a lower roton energy and provide a mechanism for roton-roton interaction. He proceeded to estimate the strength of the roton-roton interaction in order to demonstrate that such an interaction should lead to a correction of the Landau expression for the normal-fluid density, and hence should influence the calculation of the λ point. He further noted that Δ would decrease with pressure (thus providing a mechanism for phonon-roton interaction) and that the liquid would shrink if the number of rotons were increased. Feynman's arguments were well founded, and not a little prophetic. Some 25 years were to pass, however, before a more accurate theory of the consequences of roton-roton interactions was developed.[10,11]

Feynman-Cohen Wavefunction and Backflow

Feynman and his student Michael Cohen took up the problem (posed by Feynman at the end of his third paper in the series) of finding a better wavefunction to describe rotons, using as a basic building block the incorporation of backflow into the trial wavefunction.[12] As a trial problem they first considered the motion of an impurity that had the same mass m as He^4, experienced the same forces and was not subject to Bose statistics. The trial wavefunction they chose for the motion of this impurity incorporated the backflow of the surrounding liquid through a function $g(r)$:

$$\Psi_A = \Psi_0 \exp[i\mathbf{q}\cdot\mathbf{r}_A]\exp\left[i\sum_{j\neq A} g(r_j - r_A)\right], \tag{6}$$

where $g(r)$ had the dipolar form $A\mathbf{q}\cdot\mathbf{r}/r^3$, which it will possess at large distances. Using the variational principle, they determined A to be about 3.8 Å^3 (close to the classical value for perfect backflow, 3.6 Å^3) and the impurity effective mass to be 1.54 m not far from the hard-sphere value of 1.5 m. They therefore took for the excited-state wave-function a symmetrical version of equation (6):

$$\Psi_{FC} = \Psi_0 \sum_j \exp(i\mathbf{q}\cdot\mathbf{r}_j)\exp i\sum_{j\neq k} g(r_{jk}) \tag{7a}$$

$$\cong \Psi_0\left\{\sum_j \exp(i\mathbf{q}\cdot\mathbf{r}_j)\left[1+i\sum_{j\neq k} g(r_{jk})\right]\right\}, \tag{7b}$$

where $g(r)=A\mathbf{q}\cdot\mathbf{r}/r^3$. The resulting energy spectrum was by no means easy to calculate, involving as it did lengthy numerical calculations that depended on both two-body and three-body correlation functions. As may be seen in Figure 1, it led to a considerably improved value for the roton energy ($\Delta \simeq 11.5$ K).

Following their seminal work on the phonon-roton spectrum, Feynman and Cohen took up the question of measuring the excitation spectrum directly by studying the energy losses of monoenergetic neutrons scattered from liquid He II. They showed[13] that most of the scattering at a fixed angle would result from production or annihilation of a single excitation from the condensate, so that one would expect to observe narrow peaks in the spectrum of scattered neutrons, super-imposed on a broad background associated with energy losses in which two or more excitations are produced. They dealt in some detail with effects of finite temperature, instrumental resolution and line-width. Their work not only accelerated substantially the experimental effort to measure this spectrum, but was singularly successful in set-ting forth, in advance of the first experiment, the major features of the neutron scattering experiments[14] that are responsible for our present understanding of the excitation spectrum.

What is a Roton?

Feynman and Cohen further proposed a physical picture of the roton based on their trial wavefunction—namely, that it resembled a classical vortex ring of such small radius that only one atom could pass through its center, with the backflow describing a slow drift of atoms returning for another passage through the ring. Charles Aldrich and I subsequently showed that this rather poetic picture is not quite right: A roton should rather be thought of as a He^4 quasiparticle and its associated backflow field, of mass $m^* \sim 2.8\, m$, moving in an attractive potential well of energy -2 K, which originates in its coupling to the background density fluctuations of the liquid. It is the combination of changes in the strength of that momentum-dependent potential energy (which has its origin in the particle interactions responsible for the pair-correlation function) with the p^2 variation of the quasiparticle energy that is responsible for the minimum in the $E(p)$ curve.[15] It is, I believe, a measure of Feynman's openness and willingness to listen to others that although he was extremely fond of his vortex-ring picture (it figures in all his papers on the excitation spectrum), he evidenced no great hesitation in abandoning it as I discovered in a conversation with him after a colloquium I gave at Caltech in 1978. In that colloquium I had described the work that Aldrich and I had done; Feynman told me that he liked our work and that our roton picture seemed to be the right one.

Finally, when more careful integrations using the Feynman-Cohen wavefunction are carried out, and a sign error in one of their integrals is corrected, their results for the short-wavelength spectrum turn out to be less accurate than they reported. The mistake was discovered some 27 years later by Efstratios Manousakis, who was then a graduate student at the University of Illinois, Urbana-Champaign, working with Vijay Pandharipande on an improved variational wavefunction for excitations in liquid He^4. As figure 1 shows, the corrected[16] Feynman-Cohen result is sufficiently far off the mark that one cannot help but wonder if Feynman had known of the error, whether he might not have continued his work on rotons in an effort to improve that spectrum.

Feynman on Superfluid Flow

In an article in *Progress in Low Temperature Physics*,[17] Feynman discussed in some detail the nature of superfluid flow, with particular attention to the quantization of circulation for rotational flow, the response of the superfluid in a bucket that is rotated, and the existence of a critical velocity for superfluid flow. He showed how the formation of the quantized vortex lines, suggested by Lars Onsager in 1949 (in a paper that Feynman read only after completing his own, independent work), could enable the superfluid to respond to rotation as though it were a solid body. The quantization of circulation takes the form

$$\oint \mathbf{v}_s \cdot d\mathbf{l} = nh/m,$$

where the line integral is to be carried out over any closed path. By considering the kinetic energy contained in the vortex lines, Onsager and Feynman showed that when the superfluid is set in rotation, it would be energetically favorable to form singly quantized vortex lines in such a way as to mimic the solid-body rotation.

Feynman then considered the role that vortex lines play in more general fluid motions, with particular attention to the possibility that the critical velocity for superfluid flow could be determined by the production of vorticity. By estimating the energy required to create vortices for a typical experimental situation, he arrived at a quite reasonable estimate of the superfluid flow velocity (~ 1 m/sec). He then made a pioneering effort to consider the nature of superfluid turbulence. The paper is a classic: It addresses a range of extremely difficult problems in an imaginative, very physical and correct way; and it set the stage for all future work on superfluid flow.

Feynman on Superconductivity

Feynman worked very hard on superconductivity during 1954–57. I was aware of this in a general way from reading references to superconductivity in several of his papers written during this period and from listening to his 1956 lecture to the International Congress on Theoretical Physics in Seattle. However, the extent to which this problem had become a central focus for his research became clear to me only when I visited Caltech for two weeks in December 1956. Feynman and I had lunch together nearly every day; walking to lunch, at lunch and after lunch, Feynman would try out a new set of ideas about superconductivity on me. (An incomplete list of these ideas may be found in his published Seattle lecture.[18]) It was clear from these conversations that he was thoroughly familiar with both the experimental and the theoretical literature on superconductivity. Independently of Arkady Migdal, Feynman had tried his diagrammatic techniques on the electron-phonon system and found that the system continued to behave normally. Cohen pointed out to him that this was perhaps not surprising, since the true ground state might not be obtainable using perturbation theory—a quite prophetic remark.

Feynman's strategy for solving superconductivity, as described in his 1956 lecture, is worth recounting:

> I would like to maintain a philosophy about this problem which is a little different from usual: It does not make any difference what we explain, as long as we explain some property correctly from first principles. If we start honestly from first principles and make a deduction that such and such a property exists—some property that is different for superconductors than for normal conductors, of course—then undoubtedly we have our hand on the tail of the tiger because we have got the mechanism of at least one of the properties. If we have it correct we have the clue to the other properties, so it isn't very important which property we explain. Therefore, in making this attempt, the first thing to do is to choose the easiest property to handle with the kind of mathematics that is involved in the Schrödinger equation. I want to summarize some thoughts on this question, although they do not represent a solution. They represent a statement of the problem and a little bit of a personal view.
>
> I decided it would be easiest to explain the specific heat rather than the electrical properties ... . But we do not have to explain the entire specific heat curve; we only have to explain any feature of it, like the existence of a transition, or that the specific heat near absolute zero is less than proportional to T. I chose the latter because being near absolute zero is a much simpler situation than being at any finite temperature. Thus the property we should study

is this: Why does a superconductor have a specific heat less than T?

Unlike many of the other great theoretical physicists who worked long and unsuccessfully trying to develop a microscopic theory of superconductivity, Feynman was quick to recognize that John Bardeen, Leon Cooper and Robert Schrieffer had indeed solved the problem in their epochal 1957 paper. Thus at the Kamerlingh Onnes Conference in 1958 he remarked that superconductivity had been solved, and in his 1961 lectures on statistical mechanics he described the BCS theory in considerable detail.

Another indication that he retained his interest in the field is the description of the Josephson effect in the *Feynman Lectures*, which indicated that he clearly realized the implications of this work, though he never worked on this subject himself.

Feynman on Polarons

Electrons in ionic crystals are coupled (often strongly) to the nearly frequency-independent optical modes, so that as an electron moves it is accompanied by a polarization wave that acts to reduce its energy and increase its mass. The resulting entity, the electron plus its accompanying phonon cloud, is called a polaron. The polaron problem, first worked on by Landau and S. I. Pekar in the 1930s and revived by Herbert Fröhlich in the late 1940s, is of intrinsic theoretical interest because the dimensionless electron-phonon coupling constant α is typically large, so that perturbation-theoretic methods do not apply. It attracted the interest of many of us in the early 1950s because of the possibility that by understanding such a strong-coupling problem we might be able to make progress on a microscopic theory of superconductivity. Thus when I came to Urbana in 1952 to work as a postdoc with Bardeen, he suggested that I look into polarons, with the result that Tsung-dao Lee, Francis Low and I developed an intermediate-coupling theory of polarons at the University of Illinois, Urbana-Champaign, while quite similar work was carried out independently by Fröhlich and his students in Liverpool.

Feynman became interested because he saw the polarons as an opportunity to test the power of his path integral approach by using a variational principle to compute the ground-state energy for the electron coupled to the phonon field. He was right; with comparatively little effort, he was able to reproduce our intermediate-coupling results and to obtain, by a different choice of parameters in the trial actions in his path integral, accurate results for the ground-state energy and effective mass that extend smoothly into the strong-coupling domain.[19]

(Incidentally, Bardeen's intuition that understanding polarons might help in the development of a microscopic theory of superconductivity proved to be correct. Schrieffer arrived at the BCS variational wavefunction by adapting the intermediate-coupling ground-state wavefunction, which Lee, Low and I had obtained, to the model Hamiltonian Bardeen, Cooper and Schrieffer had used to describe superconductivity.)

Feynman's interest in polarons continued into the 1960s. With his postdoc Robert Hellwarth and his students Carl Iddings and Philip Platzman, he extended his variational path integral approach to the calculation of the polaron response to an external field (in unpublished calculations, Feynman had independently arrived at the Kubo-Lax description of transport properties in terms of response functions), and so obtained an expression for the polaron mobility for arbitrary coupling constant. This further enabled them to examine transport under conditions for which the Boltzmann equation is not an adequate approximation.[20]

The Feynman Legacy

Following the seminal work of Migdal and his students Spartak Beliaev and Victor Galitskii, and of Murray Gell-Mann and Keith Brueckner, in the mid-1950s, Feynman diagrams have become one of the major methods used to calculate and describe physical processes in condensed matter physics. Yet they were almost never so used by Feynman. This seeming paradox is resolved when one realizes that the problems in condensed matter physics that interested Feynman were those of strongly interacting systems, for which a variational approach, either by itself or combined with his path integral formulation, was ideally suited. Summing Feynman diagrams offers a marvelously compact way of carrying out consistent perturbation-theory calculations and, in some cases, of giving general proofs. Thus for the problems that interested Feynman, diagrams were therefore not very useful, and Feynman made only sparing use of them in his papers on condensed matter physics.

For a number of years, Feynman spent a day each week at Hughes

Research Laboratories; during this period he gave a series of lectures on topics that interested him, as well as consulted with colleagues there. In 1961 he lectured on statistical mechanics; the notes on his lectures, which were taken by R. Kikuchi and H. A. Feiveson, rapidly became a *samizdat* classic in the field. Some years later, when the publisher W. A. Benjamin approached Feynman seeking to publish these notes, he told me that given that some time had passed and he was no longer working in the field, he could no longer be sure that they would be of interest; he would therefore publish only on the condition that I first read them through and guaranteed their continuing interest. This I did, and had no difficulty in providing the requisite guarantee because the notes contain a remarkably lucid and at times quite personal exposition of statistical mechanics. On reading them one realizes anew what excellent taste Feynman had in identifying those topics that would best provide the beginner with insight into the key methods and concepts of statistical mechanics, and how willing Feynman was to explore a topic in depth, from its beginnings to current research. By now generations of graduate students (the book is in its 11th printing) have come to share this view, and I have little doubt that these notes[21] will join his other lecture notes and books as part of the library of the working physicist for generations to come.

As with the other fields in which Feynman worked, his influence on condensed matter physics was profound, and will continue to be so. Feynman diagrams and path integrals have become indispensable tools for theorists and experimenters alike, and backflow is now a significant part of the vocabulary of the physicist studying many-body problems. His work on polarons represents a difficult, if not impossible, act to follow, while his papers on liquid helium set clearly the signposts for future work in the field, work that will incorporate at every level the physical picture he set forth some 35 years ago.

References

1. F. London, Phys. Rev. **54**, 947 (1938). For a further development of London's seminal ideas, see also F. London, *Superfluids*, vol. 2, Dover, New York (1954).
2. L. Tisza, Nature **141**, 913 (1938); C. R. Acad. Sci. **207**, 1035, 1186 (1938); Phys. Rev. **72**, 838 (1947).
3. L. D. Landau, J. Phys. USSR **5**, 71 (1941); Phys. Rev. **60**, 354 (1941); J. Phys. USSR **8**, 1 (1944); J. Phys. USSR **11**, 91 (1947).
4. R. P. Feynman, Phys. Rev. **91**, 1291 (1953).

5. R. P. Feynman, Phys. Rev. **91**, 1301 (1953).
6. R. P. Feynman, Rev. Mod. Phys. **20**, 367 (1948).
7. D. M. Ceperley, E. L. Pollock, Phys. Rev. Lett. **56**, 351 (1986).
8. R. P. Feynman, Phys. Rev. **94**, 262 (1954).
9. As Feynman notes, similar wavefunctions had been proposed earlier, for example, by A. Bijl, Physica **7**, 896 (1940).
10. K. S. Bedell, I. Fomin, D. Pines, J. Low Temp. Phys. **48**, 417 (1982).
11. K. S. Bedell, A. Zawadowski, D. Pines, Phys. Rev. B **29**, 102 (1984).
12. R. P. Feynman, M. Cohen, Phys. Rev. **102**, 1189 (1956).
13. M. Cohen, R. P. Feynman, Phys. Rev. **107**, 13 (1957).
14. The pioneering experiments were carried out by H. Palevsky and his collaborators at Brookhaven in 1957. For reviews, see A. D. B. Woods, R. A. Cowley, Rep. Prog. Phys. **36**, 1135 (1973); D. L. Price, in *Physics of Liquid and Solid Helium*, vol. 2, K. H. Bennemann, J. B. Ketterson, eds., Wiley, New York (1978), p. 675.
15. For a review of this approach, see D. Pines, Can. J. Phys. **65**, 1357 (1987).
16. E. Manousakis, V. R. Pandharipande, Phys. Rev. B **30**, 5062 (1984).
17. R. P. Feynman, in *Progress in Low Temperature Physics*, vol. 2, C. J. Gorter, ed., North-Holland, New York (1955), p. 17.
18. R. P. Feynman, Rev. Mod. Phys. **29**, 205 (1957).
19. R. P. Feynman, Phys. Rev. **97**, 660 (1955).
20. R. P. Feynman, R. W. Hellwarth, C. K. Iddings, P. M. Platzman, Phys. Rev. **127**, 1004 (1962).
21. R. P. Feynman, *Statistical Mechanics*, Addison-Wesley, Reading, Mass. (1972).

This article was originally published in the February 1989 issue of *Physics Today*.

The Teacher at Cal Tech

RICHARD P. FEYNMAN, TEACHER

David L. Goodstein

One of the principal purposes of this article is to consider Dick Feynman in his role as teacher. Let me not keep you in suspense about my conclusion. I think Dick was a truly great teacher, perhaps the greatest of his era and ours. That's not to say he was always completely successful, as he himself emphasized in his preface to *The Feynman Lectures on Physics.*[1] I would contend that these lectures often failed at the level of their superficial intent: If his purpose in giving them was to prepare classes of adolescent boys to solve examination problems in physics, he may not have succeeded particularly well; if his purpose in creating those three red volumes was to provide effective introductory college textbooks, he may not have succeeded, either. If, however, his purpose was to illustrate, by example, how to think and reason about physics, then, by all indications, he was brilliantly successful. Perhaps this is why the books are genuine and lasting classics of the scientific literature and why his lectures left an enduring trace on those fortunate enough to have heard or read them.

Feynman's role as a teacher was somewhat unconventional, like almost everything else about the man. He loved puzzles and games. In fact, he saw all the world as a sort of game, whose progress, or "be-

David Goodstein is Vice Provost and a Professor of Physics and Applied Physics at the California Institute of Technology, in Pasadena, California.

havior," follows certain rules, some known, some unknown. Given the "known" rules, find the behavior; given the behavior, find the rules. Find places or circumstances where the rules don't work, and invent new rules that do. This attitude was a central theme in all of his teaching.

Perhaps I can shed some light on Dick's personality and approach to teaching by recounting a few personal anecdotes, incidents that haven't yet made it into the "Feynman story" literature because they're not the kinds of stories he liked to tell about himself and I was the only other witness.

The Joy of Immersion

The first story goes back to the week that I learned from his secretary, Helen Tuck, that Dick had cancer. She told me he was to go into the hospital for surgery the following week. He might not survive. This was, I think, in June 1979.

I saw him that Friday morning, while we were "robing up" for graduation (yes, Feynman put on silly academic robes and marched in the commencement procession the week before his first cancer operation). Someone had told me there was something wrong with a calculation that Dick and I had worked on together, but I couldn't find the source of the mistake. Would Dick like to talk about it? We made an appointment for the following Monday morning.

On Monday morning we got to work. Or rather, he did. I mostly watched and commented, and marveled to myself about this man, facing the abyss but working with unflagging patience and energy on an arcane problem in two-dimensional elastic theory. Of course, he didn't know that I knew his terrible secret.

The problem proved intractable—at six o'clock that evening, we hadn't succeeded. He declared the situation hopeless and went home.

Two hours later, Feynman called me at home with the solution to the problem. He was very excited. He had not been able to stop working on it and had finally solved it. This man, who was now four days away from a major operation, was in a *very* good mood.

Giving Credit

The second story goes back to the beginnings of the same collaboration in which we made the mistake I have just described. Feynman and I had been discussing some experiments that one of my students had done. One morning he marched into my office, walked to the blackboard and said, "Look, it's obvious that...," and proceeded in a few minutes to sketch out an idea that might explain our results. I was dumbstruck. It was simple, intuitive, beautiful. I got to work immediately putting the data in a form that could be compared with his model. It worked pretty well, so I wrote the first draft of a paper.[2] Just as I was finishing it, I got a preprint in the mail from two English physicists, J. Michael Kosterlitz and David Thouless. It presented exactly the same theory Feynman had sketched out on my blackboard.

In my experience, beneath the surface of every scientist there lurks a wounded person who believes his work has not been fully appreciated. Feynman was a rare, perhaps even unique exception. In fact, many times I saw him go to some length to make sure that he didn't take credit away from some younger theorist who needed it much more. When the Kosterlitz-Thouless paper arrived, I went directly to Feynman to tell him what had happened. For just an instant I saw the smallest shadow of disappointment flicker across his face. Then he brightened and said (apparently thinking that Kosterlitz and Thouless were one person), "Look, if two guys in different parts of the world, thinking about different problems, get the same idea, it must be right!" The Kosterlitz-Thouless theory went on to become one of the most important theories in statistical mechanics.

The Performer

Let me tell just one more personal story, a small one, but one that hints, I think, at an essential element of Dick's motivation to teach. In 1968 my wife and I returned to Caltech from a postdoctoral year in Italy. I was now an assistant professor, but we were in debt, owned almost nothing and had moved into an apartment near campus that had virtually no furniture at all. In fact, we had only two possessions to speak of, both acquired in Italy. One was a 2700-year-old Etruscan pottery drinking cup, and the other was a great big electric espresso machine. One day I invited Feynman and others over to the apartment for a cup of expresso.

The instant he entered the apartment, Feynman spotted the Etruscan drinking cup (not hard to do because there was almost nothing else around). He immediately picked it up and started playing with it, turning it over, tapping the surface and boasting that he would explain how you could tell it wasn't genuine. I saw my wife go absolutely white in stark terror when he bounced the cup in his hand, proceeding to tell us stories about amazing things that had been discovered in Etruscan tombs. In fact, he didn't damage the cup, and he didn't prove it was a fraud. What he did do was just what Feynman always did. He absolutely riveted the attention of everyone in the room for the entire time he was there. His need to do that helps explain some of the racy stories he liked to tell about himself, but it also lies close to the core of what made him a great teacher. For Feynman, the lecture hall was a theater, and the lecturer a performer, responsible for providing drama and fireworks as well as facts and figures. This was true regardless of his audience, whether he was talking to undergraduates or graduate students, to his colleagues or the general public.

I can remember many moments of high drama in Feynman lectures. Once, for example, several years ago, he taught a course in advanced quantum mechanics in a large lecture hall, to a class consisting of a few registered graduate students and most of the Caltech physics faculty. At one point he started explaining how to represent certain complicated integrals diagrammatically: time on this axis, space on that axis, wiggly line for this, straight line for that and ... turning around suddenly to face the class with a wicked grin he said, "and this is called *the diagram!*" The class ignited with spontaneous applause.

A more recent memorable moment comes from the last lecture I heard Feynman give, a guest lecture at Caltech's freshman physics course. These appearances had to be kept secret so there would be room for the freshmen in the hall. The subject was curved space-time, and the lecture was characteristically brilliant. But the unforgettable moment came at the beginning before he really got started. The supernova of 1987 had just been discovered, and he was very excited about it. "Tycho Brahe had his supernova," Dick said, "and Kepler had his. Then, there weren't any for 400 years. But now, I have mine." Then he went on to defuse the awed silence he had created. "There are 10^{11} stars in the galaxy," he said. "That used to be a *huge* number. But it's only a hundred billion. It's less than the national deficit! We used to call them astronomical numbers. Now we should call them economical numbers." The class dissolved in laughter, and Dick went on with his lecture.

In his public lectures Dick never hesitated to say exactly what was on his mind. When he went to speak at another university—I personally saw this happen on two separate occasions—he would express his views on psychology, and the psychology department could absolutely be depended on to rise and depart on cue, and *en masse*. Then he would do the same for philosophy. You can imagine the fiendish delight Dick took from these demonstrations. But he also knew how to illuminate a subtle point with a brilliant example that anyone could understand. I remember one time he was trying to explain why you must not verify an idea by using the same data that suggested the idea in the first place. This is a point even many scientists don't appear to understand. Seeming to change the subject, Feynman said: "You know, the most *amazing* thing happened to me tonight. I was coming here, on the way to the lecture, and I came in through the parking lot. And you won't believe what happened. I saw a car with the license plate of ARW 357! Can you imagine? Of all the millions of license plates in the state, what was the chance that I would see that particular one tonight? Amazing!"

The Official Record

I felt I should do some real research to prepare to write here about Feynman's teaching, so I decided to find out what he had actually taught during his career. I don't have any information on what courses he gave in his early days at Cornell, but I do have the record of what he taught at Caltech.

In 35 years, from 1952 to 1987, he was listed as teacher of record for 34 courses. Most of these, in fact 25 of them, were advanced graduate courses; in typical Caltech fashion these were strictly limited to graduate students, unless undergraduates asked permission to take them. (They often did and it was nearly always granted.) These courses included advanced quantum mechanics, the course he taught more than any other course (nine times), and second most often (or five times) a course called Topics in Theoretical Physics—in other words, whatever he felt like talking about. He also taught Elementary-Particle Theory and High-Energy Physics, which were separate courses during the 1960s, the heyday of these subjects at Caltech. He taught other graduate courses such as relativity, and a couple of times he taught introductory graduate courses, including mathematical methods of physics and quantum mechanics.

In 1981, toward the end of his career, Feynman joined with John Hopfield and Carver Mead to offer an interdisciplinary course called The Physics of Computation. Two years later Hopfield and Mead were still teaching The Physics of Computation, but Dick had split off a

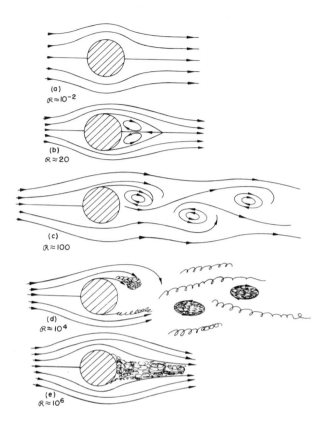

Pedagogical simplicity of the diagrams in The Feynman Lectures on Physics is exemplified by this schematic figure (Figure 41–6, Volume II) showing fluid flow past a cylinder for various Reynolds numbers.

separate course called Potentialities and Limitations of Computing Machines. I have not determined whether this represented a broadening of our offerings in this area or a schism in the high church of computing. In any case, in all those years, only twice did he teach courses purely for undergraduates. Those were the celebrated occasions in the academic years of 1961-62 and 1962-63 when he lectured, first to the freshmen, and then again to the same set of students when they were sophomores, on the materials that were to become *The Feynman Lectures on Physics.*

Based on these data, should Dick be remembered as relatively untried in the teaching of undergraduates? Not really: In spite of the formal, written record, the real story—that of informal contact with undergraduates—is a little different. For many years—at least 17, but there is no written record to check—he also taught an informal course called Physics X. This class, for which no credit was offered, met weekly, on Monday or Tuesday afternoons at 5 pm, which was the most convenient time for the students. The curriculum consisted of whatever the students felt like discussing. There was, in fact, just one unbreakable rule: No faculty were permitted to attend. As a result, I can't tell you anything else about what went on in Physics X. There is one more thing I can tell you, though. Tuck, who was Feynman's secretary for 17 years, had the job of deflecting visitors, avoiding appointments and generally protecting Feynman's privacy. There was one standing exception to this rule: The door was always, unconditionally, open to any student who wanted to see him.

The Undergraduate Lectures

Dick once told me that in the long run, his most important contribution to physics would be not QED or the theory of superfluid helium or polarons or partons. His real monument would be his *Feynman Lectures*. I think we can all agree at least that it was his most important contribution to physics education. I didn't arrive at Caltech until a few years after it was completed, so in an attempt to find out how it was all done and how it went, I've talked to a number of people who where there. Here's a bit of what I've gleaned.

It was apparently Matthew Sands who had the idea of asking Dick to give the lectures. There was a feeling that the Caltech students, who were among the best in the country, were getting turned off rather than turned on by their two years of compulsory physics, and something had to be done. When Dick agreed to do the job, it was immediately decided to transcribe the lectures and publish them. That job turned out to be far more difficult than anyone had imagined. Turning out publishable books required a tremendous amount of work on the part of Sands, Robert Leighton, Gerry Neugebauer, Rochus Vogt and many others, as well as Feynman himself.

Meanwhile, there was the matter of how to take care of the nuts and bolts of running a course for nearly 200 students. This task was greatly complicated by the fact that Feynman used only a minimal outline of what he wanted to cover, so that no one but he knew what he was going to say until he said it. A single sheet, 8 1/2 by 11, with cue words and diagrams was the norm.

This scheme was particularly hard on Neugebauer, a young assistant professor at the time, who had the job of turning out homework assignments for the students on the afternoon after each lecture. The problem, he says, is that he sometimes didn't capture all of the lecture. To help out with this problem, Feynman, Leighton, Sands and Neugebauer would have lunch together after each lecture at the campus cafeteria, known affectionately to generations of Caltech students as "the Greasy." At these sessions, Leighton, Sands and Feynman would rehash the finer points of the lecture while Neugebauer (as he tells the story) would try desperately to pick up enough additional comprehension to be able to make up homework problems.

Incidentally, when I arrived at Caltech as a postdoc five years later, Neugebauer, who was by then well launched on his distinguished career, was the lecturer in Physics 1, and I was pressed into service as a teaching assistant, teaching a recitation section. Feynman still came to lunch after each lecture, and it was at those lunches at the Greasy that I first really got to know him.

Many of the students, Gerry says, feared the course when Feynman taught it. I've spoken to some of those students in recent times, and in the gentle glow of dim memory, each has told me that having two years of physics from Feynman himself was the experience of a lifetime. But that's not the way Gerry remembers it. As the course wore on, attendance by the kids at the lectures started dropping alarmingly, but at the same time, more and more faculty and graduate students started attending, so the room stayed full, and Feynman may never have known he was losing his intended audience.

Why did Feynman accept the assignment to devote all of his formidable energy for two years to teaching beginning physics as it had never been taught before? My guess is that there were three basic reasons. One was that, as we saw earlier, he loved to have an audience, and this gave him a bigger one than he usually had in the graduate courses he normally taught. The second was that he genuinely cared about students. He simply thought that teaching freshmen was an important thing to do, and so he couldn't turn down the invitation when it came. The third reason—and this might have been the most important of all—was the sheer challenge of reformulating physics, as he understood it, so that it could be presented to young students. This was a specialty of his. In fact, it was the standard by which he measured whether something was really understood. Once I asked him to explain to me, so that I could understand it, why spin-1/2 particles obey Fermi-Dirac statistics. Gauging his audience perfectly, he said,

"I'll prepare a freshman lecture on it." But a few days later he came to me and said: "You know, I couldn't do it. I couldn't reduce it to the freshman level. That means we really don't understand it."

It's true, as we've seen earlier, that he sometimes missed the mark. The lessons in physics he prepared, the explanations of physics at the freshman level, weren't really for freshmen, but were for us, his colleagues. As I reread the red books, which I did last summer, I thought every once in a while that I caught him looking over his shoulder, not at his young audience, but directly at us, and saying: "Look at that! Look how I finessed that delicate point! Wasn't that clever?" But even when he thought he was explaining things lucidly to freshmen or sophomores, it was not always really they who benefited most from what he was doing. It was more often us, scientists, physicists, professors, who would be the main beneficiaries of his magnificent achievement, which was nothing less than to see all of physics with fresh new eyes. Feynman was more than merely a great teacher. His lasting monument is that he was a great teacher of teachers.

I will finish by quoting something directly from the source. I have my own favorite passages, of course. Probably we all do. But I'm going to choose one that he, in effect, chose himself. One day a few years ago, I ran into him on campus. He was in a state of high excitement, which for him was perfectly normal of course, but the cause of his excitement was a passage from *The Feynman Lectures*, which he was brandishing in photocopy form. "Look at this," he said. "I said this long before they sent the first mission into space." This was probably at the time of one of the Viking orbiters and landers, which were particularly exciting events at Caltech, since JPL was running the show. By then we had all learned that life was not going to be discovered in the solar system as many had originally hoped, but still the pictures of Mars coming back from the Vikings were an extraordinary experience for us all. And just as he claimed, he had, way back in 1962, anticipated just what would happen. Let me quote to you what he said. It comes from book II, chapter 41, verse 6:

> There are those who are going to be disappointed when no life is found on other planets. Not I—I want to be reminded and delighted and surprised once again, through interplanetary exploration, with the infinite variety and novelty of phenomena that can be generated from such simple principles. The test of science is its ability to predict. Had

you never visited the Earth, could you predict the thunder-storms, the volcanos, the ocean waves, the auroras, and the colorful sunset? A salutary lesson it will be when we learn of all that goes on on each of those dead planets—those eight or ten balls, each agglomerated from the same dust cloud and each obeying exactly the same laws of physics.

References

1. R. P. Feynman, R. B. Leighton, M. Sands, *The Feynman Lectures on Physics*, three volumes, Addison-Wesley, Reading, Mass. (1963, 1964, 1965).
2. R. L. Elgin, D. L. Goodstein, Phys. Rev. A **9**, 2657 (1974).

This article was originally published in the February 1989 issue of *Physics Today*.

IT NEVER PASSED *HIM* BY

Michael Cohen

===

I entered Cornell as a freshman in 1947 and immediately heard of Bethe and Feynman. Feynman was, of course, already famous as the young genius of the campus and tales of his various exploits were widely circulated. Someone pointed him out to me in a large room in Willard Straight Hall (the student union) where people drank coffee in the morning and beer later in the day. It was said that he hung around there and used to strike up conversations with pretty girls struggling with physics homework. (He told me a few years later that he decided to leave Cornell when he tried that routine on a coed and she said, "I know who you are. You're not a student, you're Dick Feynman.")

My classmate Irwin Shapiro, who had attended the same elementary school as Feynman in Far Rockaway, was proud of the fact that he had solved some plane geometry problem in less time than Feynman. Apparently the teacher kept records!

I did not really get to know him at Cornell and did not take any of his courses, though I took courses from Mark Kac and Philip Morrison. In the Fall of 1948 I was admitted to membership in Telluride House, where I lived until I graduated in Spring 1951. This was a place where very bright (supposedly) people and student leaders lived, the cost be-

Michael Cohen is Professor of Physics at the University of Pennsylvania, Philadelphia, Pennsylvania.

ing borne by an endowment, Telluride Association which had been established in 1912 by an idealistic businessman named L. L. Nunn. Every year one or more faculty guests lived at Telluride House, and during 1947-48, the year before I first lived there, Feynman was a faculty guest. Everyone was still talking about him, and he occasionally visited. I remember he was there for a house party weekend, and once he came as an official guest at Sunday dinner. The boys at Telluride took themselves quite seriously and tried to maintain a high standard of intellectual intercourse. On Sunday several faculty would be invited for dinner. We were supposed to engage in serious discussion with them, and a House member was assigned to hover over each faculty guest to assure that someone was always talking to him or her. When Dick arrived at the house, he asked loudly "Who has me?"

The year Dick was at Telluride was when he developed most of his ideas on QED. A Telluride friend of mine, Bob Gatje, wrote Dick a letter of congratulation when he got the Nobel prize. He showed me Dick's gracious response, which appeared to acknowledge his indebtedness to the stimulating intellectual environment at Telluride. Since I knew that Dick had had a lot of fun at Telluride House, but didn't subscribe to the attitude of high intellectual seriousness, I kidded him several years later about his note to Gatje. Dick said that "Gatje didn't read my letter carefully enough. I didn't say that I did my best work because of the atmosphere at Telluride. I said, 'Could it be that the atmosphere at Telluride stimulated me to do my best work?'"

In the summer of 1950, the end of my junior year at Cornell, Mark Kac invited me to spend the summer as his research assistant at the Institute for Numerical Analysis in Los Angeles on the UCLA campus. The INA was the western division of the National Applied Mathematics Laboratory, part of the National Bureau of Standards. The list of residents that summer included not only Kac and Feynman but a whole galaxy of mathematicians, among whom the legendary Paul Erdös is the most easily remembered. My main recollection of Feynman was his brilliance in a class in which we were learning how to program for a new computer which was being built at INA, and his enthusiasm as a participant in a Saturday afternoon water polo game which we played in the UCLA pool.

The work I did with Kac consisted of taking random walks (on a computer) with a Gaussian measure, and evaluating the time integral of the square of the displacement along each path.[1] From this you could get the lowest eigenvalue of Schrodinger's equation for the harmonic oscillator potential. In his autobiography[2] Kac states that he learned about the connection between "path integrals" and partial differential equations from Feynman. I continued the random walk work myself the following summer, inventing a biased sampling procedure to pick out the most important paths. In the winter of 1950 when I was a senior at Cornell and doing some individual study with Kac, I partially reinvented Feynman's path integral approach to quantum mechanics, which was unknown to me but was obviously similar to the work I had done with Kac!

I entered Cal Tech in Fall 1951 and vaguely hoped that I would be able to work with Feynman, though I really hadn't focused on that objective. During my first year he was on leave in Brazil (he mentions in his book[3] that Bacher sweetened the package by giving him a leave his first year there). Early in my second year, Feynman walked into my office and said that it would be nice if I could help him with some integrals associated with some of his diagrams in QED. He wanted to see if you could make some kind of partial summation of a whole family of diagrams, or maybe he was studying the high energy limit (I forget). I think he indicated that Kac had told him to look me up. Needless to say, I was delighted.

Obviously Feynman didn't need my help, but was being nice. I did a lot of integrals and nothing much came of it. I remember that the integrals were obviously symmetric under the exchange of certain symbols, but our method of evaluation lost this obvious symmetry. I think Feynman was pleased that I recognized and struggled with this difficulty.

It was clear to me by this time (1953 or possibly 1954) that it would be great to keep working with Feynman and that it wasn't going to be in the area of QED. I also knew him well enough by now to recognize that he was very helpful and patient with students who exhibited some intellectual curiosity and were actually struggling with a problem and sought his advice, and was totally uninterested in dealing with people who didn't exhibit some intellectual initiative and simply wanted his sponsorship. Rather than ask him to suggest something for me to work on outside QED, I simply read his recent papers on helium and looked for places where one could go further than he had gone.

The first thing I tried took maybe several months and was a blind alley, at least for me. I tried to improve on his analysis of the lambda transition.[4] His calculation, based on the approximate counting of paths on a lattice subject to certain geometric restrictions, led to a phase transition identical with that in the ideal Bose gas. This gives a specific heat which is finite and continuous, with a discontinuous slope at the transition temperature, in contrast with the observed log-arithmic singularity (log of the magnitude of $T-T_c$) in the specific heat of He⁴. I tried to fix this up by counting paths more accurately, but I found that the main effect of improved counting was to change the transition temperature rather than the nature of the singularity.

When I saw that the lambda transition work wasn't leading any-where, I took a careful look at his work on the energy-vs-momentum relation for the elementary excitations in liquid helium. With a quite simple wave function he had obtained a curve $E(k)$ of the type postu-lated by Landau. The curve was linear for small k, corresponding to a quantized sound wave (phonon) and then dipped down to a local min-imum for k near 2 inverse angstroms. Excitations near this minimum seemed to correspond to Landau's "rotons," but there was nothing in Feynman's wave function which associated a rotational character with these states. Although Landau had suggested that these states repre-sented some kind of quantized rotational motion, there was no evi-dence (and still isn't any) for this suggestion; more importantly, Landau had estimated (quite accurately, from thermodynamic data) the energy of a minimum energy roton, and the simple Feynman wave function gave an energy which was almost twice too high.

I saw the possibility of improvement here, and it was clear that enough work would be required to make an acceptable thesis. Feyn-man's wave function for the "roton" was just the ground state wave function (which incorporated the basic restriction that the particles mustn't get too near each other) multiplied by a factor which repre-sents a single particle moving freely through the fluid (except, because of the indistinguishability of the particles, you don't know which is the moving particle; this makes the factor a *sum* of plane waves for all the particles rather than a single plane wave). I thought that the energy of the state might be lowered by requiring the particles in the neighbor-hood of the moving particle to execute some kind of evasive action. I discussed with Feynman how to put this idea into the wave function. It was clear that additional correlation factors had to be inserted. Feyn-man convinced me that the important thing was not to change the magnitude of the two-body spatial correlations, but to put in a complex correlation factor of modulus unity, which incorporates the idea that particles near a moving particle are exhibiting an orderly pattern of "backflow" to get out from in front of the moving particles and to fill in the space behind it. We agreed pretty fast on the appropriate wave function, but it looked as though the calculations might be prohibi-tively complicated. We only had hand computing facilities readily available, and I think we both took it as given that we would not trust the answers unless we could stay very close to the numbers at every stage. By introducing Fourier transforms appropriately, I showed that the calculation could be reduced to a quite manageable form, even on a hand calculator, and the project was underway. It turned out to be very successful.[5]

Feynman was a scrupulously conscientious thesis advisor. He was interested in all the details. When I finally gave him the thesis, he closeted himself with it and found an important numerical error. If the error had been undetected, we would have obtained Landau's value of the energy almost perfectly. Our "honest" result was 20% higher than the Landau value.

After the thesis I stayed on for 18 months until June 1957. In 1956 I got myself a summer job at Brookhaven National Lab where Feynman was also a visitor. Lee and Yang were there, thinking about parity nonconservation, and the intellectual temperature was quite high. Feynman and I had already talked about the possibility of direct measurement of the energy-vs-momentum curve for the excitations by producing them by scattering slow neutrons from the liquid. If you could measure the energy distribution of neutrons which had experienced a given momentum change, a sharp line in the distribution would be produced by neutrons which had produced a single excitation. The difference in energy between the incident neutrons and the neutrons in this line would be the energy of the excitation.

We worked out the detailed theory of the proposed neutron experiment.[6] Our most important theoretical prediction was that a very large fraction (between one-half and one) of the total neutron-scattering cross section would go into the production of single excitations. Therefore the experiment was feasible. Harry Palevsky, a neutron physicist at Brookhaven, discussed the experiment with us in much detail and did it with Otnes and Larsson in Sweden while on leave in 1957.[7] The experiment was subsequently done by Yarnell et al.[8] at Los Alamos, and many others. This provided striking verification of the existence of elementary excitations with a well-defined energy-vs-momentum relation of the form postulated by Landau and calculated by us.

I stayed with Feynman until spring 1957. When Oppenheimer visited Pasadena in 1957, Dick recommended to him that he make me a member of his Institute for the following year, which Oppie did. Every minute with Dick was fun. He didn't psychoanalyze me, and I will reciprocate.

Feynman was much amused that I was taking lessons at the Arthur

Murray Dance Studio in Pasadena, since he was a superb self-taught dancer and found it remarkable that I would have to spend money to learn to dance. When I would come to his office to discuss physics, he would ask for a progress report from Arthur Murray, and I tried always to have an amusing tale for him. The one he liked best was the account of how I was dancing with my instructor, stepping on her feet, when the studio manager came by with an elderly couple whom he was setting up for a "life membership" (a huge amount of money). The manager explained "This is Mr. Cohen, one of our beginning students." I said "It's not true. I've had 1000 hours of lessons" and then proceeded to dance even worse than usual. Dick's *chutzpah* was contagious.

Another time, he insisted that I demonstrate the tango, which I had just "learned." I loped into the step like Groucho Marx, but ran into trouble when I reached the wall of Dick's office. He asked "Didn't they teach you how to turn?" We decided that they had taught me how to dance in a room with periodic boundary conditions: when you bump into a wall you are assumed to reappear at the opposite wall, going in the same direction!

I once went with him and a mutual friend, Dave Elliott, to Las Vegas for a weekend. He had endless patience in the quest for women, and said it was necessary to sit at a bar for a very long time in order to attract their attention without appearing too eager. He was pleased when an attractive woman warmed up to him, and became annoyed when he found that I had told her (in his absence) that he was the smartest man in America; he wanted them to like him for his personality, not because he was important. Elliott and I kidded him that his personality wasn't that special, at least in the eyes of women at bars, and that only physicists really appreciated his act. I also won quite a bit of money from him in a gambling game which we invented, based on what came up on the roulette wheel.

One afternoon a fellow graduate student, Don DuBois, gave a seminar on some very painstaking work he had done, extending Gell-Mann and Brueckner's work on the electron gas.[9,10] Before Don started, he temporarily left his notes on a seat next to Dick and me. We noticed some incredibly complicated integral which he evaluated in several pages, getting the arctan of the square root of some mess. When this integral came up in Don's lecture, Dick said "That's easy. It's the arctan of ... ." Don was thunderstruck.

I recall sitting next to Dick at another theoretical talk shortly before I left Cal Tech. While waiting for the talk to start, I remarked that it would be interesting if you could pinpoint the exact instant in time when physics passes you by. It was a hot afternoon, and after a while I fell asleep. Suddenly he shook me and said "There it goes!"

It never passed *him* by.

References

1. M. Kac and M. Cohen, National Bureau of Standards Report No. 1553, 1952. This work was a Stone Age predecessor of modern Monte Carlo calculations of lattice gauge theories.
2. M. Kac, *Enigmas of Chance* (Harper and Row, New York, 1985), p. 116.
3. R. P. Feynman, *Surely You're Joking, Mr. Feynman* (Norton, New York, 1985), p. 233.
4. R. P. Feynman, Phys. Rev. **91**, 1291 (1953).
5. R. P. Feynman and M. Cohen, Phys. Rev. **102**, 1189 (1956).
6. M. Cohen and R. P. Feynman, Phys. Rev. **107**, 13 (1957).
7. H. Palevsky, K. Otnes, and K. E. Larsson, Phys. Rev. **112**, 11 (1958).
8. J. L. Yarnell, G. P. Arnold, P. J. Bendt, and E. C. Kerr, Phys. Rev. **113**, 1379 (1959).
9. M. Gell-Mann and K. A. Brueckner, Phys. Rev. **106**, 364 (1957).
10. D. F. DuBois, Ann. Phys. (New York) **7**, 174 (1959).

REFLECTIONS ON DICK FEYNMAN AS AN ACOLYTE AND AS HIS BOSS

Marvin L. Goldberger

I first heard about Dick while I was working on the Manhattan Project. He and Julian Schwinger were described as the emerging geniuses of theoretical physics, both having shown unusual precociousness before and during the war. Our first actual meeting took place at an American Physical Society meeting in Washington in the spring of 1947. I happened upon him in Rock Creek Park where he began holding forth to a very small group his ideas about positrons being viewed as electrons running backward in time and of very powerful computational methods in the field of quantum electrodynamics that enormously simplified, what were by traditional methods, horrendous tasks. He didn't have at that time any very convincing derivation of these techniques but he was convinced that he was onto something and he certainly bowled over this group of graduate students who were hanging onto his every word and were swept away by his enthusiasm. Perhaps more sank in than I realized, because in 1948 Bob Serber lectured to us at Berkeley what he had understood from Feynman, on

Marvin L. Goldberger is at the University of California at San Diego. Before that, he was the Director of the Institute for Advanced Study in Princeton and President of the California Institute of Technology.

a visit to Cornell, and it all seemed very simple and natural. Even after Dyson had shown how one could derive the Feynman rules from the conventional formalism of field theory, I have always approached these calculations from Feynman's standpoint (as taught to me by Serber).

In the spring of 1949 while I was a post doc at the Radiation Lab at Berkeley, I was asked to give an invited paper at a Physical Society meeting in Seattle. I talked about some work I had been doing on photopion production. Dick was also an invited speaker, talking about quantum electrodynamics. We spent a whole afternoon together talking about field theory. I was enormously flattered to have his undivided attention. He expressed amazement that I had found it completely straightforward to use his methods to actually compute cross sections with all the factors of 2π in place. I have no detailed recollection of our conversation, only that I enjoyed it immensely and found him totally without pretentiousness.

In the following nearly thirty years I saw little of Dick except at high energy physics conferences (we did have an amusing encounter in 1955 to which I'll return below), where he was a lively presence. I remember in particular two remarks he made to me at the 1958 "Rochester" conference which was held in Geneva. I had given a talk about what has come to be called the Goldberger-Treiman relation, which showed a quite unexpected connection between weak and strong interactions. Our derivation of the result was based on a number of rather questionable assumptions (of which we were quite aware). Dick came up afterward and said "your derivation stinks but the result looks right and is very important." As we went off to lunch, his second remark was that "so-and-so," an old intellectual competitor, spoke so clearly that you could tell immediately that he had nothing to say. That was the only nasty comment (and was said really in fun because I know he had great respect for the individual) I ever heard him make about a person. He could be merciless at a colloquium on a question of physics but not in an ad hominum fashion. One other thing I remember about that Geneva visit was being taken by Dick to the leading night club, the Bataclan.

In 1955 I was on the faculty of the University of Chicago. Fermi had died the previous year and the physics department decided we should try to hire Feynman. I was dispatched along with the Dean of the Faculty to make the case. The Dean, Walter Bartky (a distinguished applied mathematician) was afraid to fly, so we set out on the Super Chief for Pasadena. We got off the train and took a taxi to Feynman's house. We talked for a while about the possibility of his coming to Chicago. He was utterly uninterested and refused to even listen to what was an extravagant salary. He had essentially locked his wife (not Gweneth, a previous one) into a closet because, I guess, he thought she might be tempted by the offer. Dick's position was that he had made one difficult decision in leaving Cornell, and feeling that it was a correct one didn't want to waste time ever again agonizing over alternatives. Frustrated and disappointed we got back on the train that afternoon and returned to Chicago.

The story picks up again when I went to Cal Tech as president. Dick was in the hospital when I arrived, recovering from a horrendous operation for the removal of a four kilogram malignant kidney tumor. I went to see him and the first thing he said to me was that he would rather be where he was than where I was. The second thing he said was that he wouldn't do anything I wanted him to do at Cal Tech. He actually relented on this on a couple of occasions. He was vitally interested in the intellectual aspects of being a faculty member, especially in teaching and participating in seminars, but he had no patience for committees or faculty politics. He had very few Ph.D. students and was less responsible than I think he should have been for the welfare of his younger colleagues. I have no knowledge of whether he didn't like to have collaborators or whether he only rarely found ones that were congenial. The fact is that in physics he was very much a loner.

After the war, to the best of my knowledge, Dick never was involved with classified research or any of the government science advisory apparatus. He gave the impression of being totally apolitical. In 1982, however, he became sufficiently incensed by Reagan's military posture and refusal to pursue arms control that he began speaking out in support of the nuclear freeze movement which was quite strong in California. Needless to say, that given his stature and his eloquence, he was very effective. His participation on the Rogers committee investigating the Challenger disaster is well known; his willingness to involve himself in that surprised me. The country owes a great deal to his effectiveness in that task.

Dick had a rare ability to be able to concentrate the full force of his incredible intelligence on **your** problem when you went to him with a question. Fermi was the only other person I've ever encountered with this attribute. It didn't matter what the subject was; everything was a challenge to be understood, and usually in a totally unexpected way. He approached problems with the attitude of a brilliant child unincumbered with inhibitions of previous knowledge. I remember once discussing with Schwinger some work I was doing which was an extension of an amusing thing I had learned from Dick. After explaining to Julian what Dick had done he said: "That Feynman, he always sees something no one else has noticed."

There is no question that Dick was an egotist and a terrible show off. His pleasure in what he did was so great and spontaneous that I, at least, didn't find it offensive, even if occasionally boring. The thing, however, that endeared him most to me was that he felt no need to put down other people or their accomplishments.

His courage and spirit during the last few years of his life were an inspiration for all of us who had the privilege of knowing him.

The Public Physicist
and Consultant

RICHARD FEYNMAN AND THE CONNECTION MACHINE

W. Daniel Hillis

One day in the spring of 1983, when I was having lunch with Richard Feynman, I mentioned to him that I was planning to start a company to build a parallel computer with a million processors. (I was at the time a graduate student at the MIT Artificial Intelligence Lab). His reaction was unequivocal: "That is positively the dopiest idea I ever heard." For Richard a crazy idea was an opportunity to prove it wrong—or prove it right. Either way, he was interested. By the end of lunch he had agreed to spend the summer working at the company.

Richard had as much fun with computers as anyone I ever knew. His interest in computing went back to his days at Los Alamos, where he supervised the "computers," that is, the people who operated the mechanical calculators. There he was instrumental in setting up some of the first plug-programmable tabulating machines for physical simulation. His interest in the field was heightened in the late 1970s when his son Carl began studying computers at MIT.

I got to know Richard through his son. Carl was one of the undergraduates helping me with my thesis project. I was trying to design a

W. Daniel Hillis is the founder of Thinking Machines Corporation in Cambridge, Massachusetts, and the inventor of the firm's Connection Machine.

computer fast enough to solve commonsense reasoning problems. The machine, as we envisioned it, would include a million tiny computers, all connected by a communications network. We called it the Connection Machine. Richard, always interested in his son's activities, followed the project closely. He was skeptical about the idea, but whenever we met at a conference or during my visits to Caltech, we would stay up until the early hours of the morning discussing details of the planned machine. Our lunchtime meeting on that spring day in 1983 was the first time he ever seemed to believe we were really going to try to build it.

Richard arrived in Boston the day after the company was incorporated. We had been busy raising the money, finding a place to rent, issuing stock and so on. We had found an old mansion just outside the city, and when Richard showed up we were still recovering from the shock of having the first few million dollars in the bank. No one had thought about anything technical for months. We were arguing about what the name of the company should be when Richard walked in, saluted and said, "Richard Feynman reporting for duty. OK, boss, what's my assignment?"

The assembled group of not-quite-graduated MIT students was astounded. After a hurried private discussion ("I don't know, you hired him..."), we informed Richard that his assignment would be to advise on the application of parallel processing to scientific problems. "That sounds like a bunch of baloney," he said. "Give me something real to do."

So we sent him out to buy some office supplies. While he was gone, we decided that the part of the machine we were most worried about was the router that delivered messages from one processor to another. We were not entirely sure that our planned design would work. When Richard returned from buying pencils, we gave him the assignment of analyzing the router.

The Machine

The router of the Connection Machine was the part of the hardware that allowed the processors to communicate. It was a complicated object; by comparison, the processors themselves were straightforward. Connecting a separate wire between every pair of processors was totally impractical; a million processors would require 10^{12} wires. Instead, we planned to connect the processors in the pattern of a

20-dimensional hypercube, so that each processor would only need to talk directly to 20 others. Because many processors had to communicate simultaneously, many messages would contend for the same wire. The router's job was to find a free path through this 20-dimensional traffic jam or, if it couldn't, to hold the message in a buffer until a path became free. Our question to Feynman was: Had we allowed enough buffers for the router to operate efficiently?

In those first few months Richard began studying the router circuit diagrams as if they were objects of nature. He was willing to listen to explanations of how and why things worked a certain way, but fundamentally he preferred to figure everything out himself. He would sit in the woods behind the mansion and simulate the action of each circuit with pencil and paper.

Meanwhile, the rest of us, happy to have found something to keep Richard occupied, went about the business of ordering the furniture and computers, hiring the first engineers and arranging for the Defense Advanced Research Projects Agency to pay for the development of the first prototype. Richard did a remarkable job of focusing on his "assignment," stopping only occasionally to help wire the computer room, set up the machine shop, shake hands with the investors, install the telephones and cheerfully remind us of how crazy we all were. When we finally picked the name of the company, Thinking Machines Corporation, Richard was delighted. "That's good. Now I don't have to explain to people that I work with a bunch of loonies. I can just tell them the name of the company."

The technical side of the project was definitely stretching our capacities. We had decided to simplify things by starting with only 64 000 processors, but even then the amount of work to be done was overwhelming. We had to design our own silicon integrated circuits, with processors and a router. We also had to invent packaging and cooling mechanisms, write compilers and assemblers, devise ways of testing processors simultaneously and so on. Even simple problems like wiring the boards together took on a whole new meaning when you were working with tens of thousands of processors. In retrospect, if we had had any understanding of how complicated the project was going to be, we would never have started.

'Get These Guys Organized'

I had never managed a large group before, and I was clearly in over my head. Richard volunteered to help out. "We've got to get these guys organized," he told me. "Let me tell you how we did it at Los Alamos."

It seems that every great man has a certain time and place in his life that he takes as a reference point ever after: a time when things worked as they were supposed to and great deeds were accomplished. For Richard, that time was at Los Alamos during the Manhattan Project. Whenever things got "cockeyed," Richard would look back and try to understand how now was different from then. Using this formula, Richard decided we should pick an expert in each area of importance to the machine—software, packaging, electronics and so on—to become the "group leader" of that area, just as it had been at Los Alamos.

Part two of Feynman's "Let's Get Organized" campaign was a regular seminar series of invited speakers who might suggest interesting uses for our machine. Richard's idea was that we should concentrate on people with new applications, because they would be less conservative about what kind of computer they would use. For our first seminar he invited John Hopfield, a friend of his from Caltech, to give us a talk on his scheme for building neural networks. In 1983, studying neural networks was about as fashionable as studying ESP, so some people considered Hopfield a little crazy. Richard was certain he would fit right in at Thinking Machines.

What Hopfield had invented was a way of constructing an "associative memory," a device for remembering patterns (see the article, "Statistical Mechanics of Neural Network," by Haim Sompolinsky, PHYSICS TODAY, December, 1988, page 70). To use an associative memory, one trains it on a series of patterns—for example, pictures of letters of the alphabet. Later, when the memory is shown a new pattern, it is able to recall a similar pattern it has seen in the past. A new picture of the letter A will "remind" the memory of another A it has seen before. Hopfield figured out how such a memory could be built from devices functionally similar to biological neurons.

Not only did Hopfield's method seem to work; it seemed to work particularly well on the Connection Machine. Feynman figured out the details of how to use one processor to simulate each of Hopfield's neurons, with the strength of each connection represented as a number in the processor's memory. Because of the parallel nature of Hopfield's algorithm, all the processors could be used concurrently with

100 percent efficiency; the Connection Machine would thus be hundreds of times faster than any conventional computer.

An Algorithm for Logarithms

Feynman worked out in some detail the program for computing Hopfield's network on the Connection Machine. The part that he was proudest of was the subroutine for computing a logarithm. I mention it here not only because it is a clever algorithm, but also because it is a specific contribution Richard made to the mainstream of computer science. He had invented it at Los Alamos.

Consider the problem of finding the logarithm of a fractional number between 1 and 2. (The algorithm can be generalized without too much difficulty.) Feynman observed that any such number can be uniquely represented as a product of numbers of the form $1+2^{-k}$, where k is an integer. Testing for the presence of each of these factors in a binary representation is simply a matter of a shift and a subtraction. Once the factors are determined, the logarithm can be computed by adding together the precomputed logarithms of the factors. The algorithm fit the Connection Machine especially well because the small table of the logarithms of $1+2^{-k}$ could be shared by all the processors. The entire computation took less time than doing a division.

Concentrating on the algorithm for a basic arithmetic operation was typical of Richard's approach. He loved the details. In studying the router he paid attention to the action of each individual gate, and in writing the program he insisted on understanding the implementation of every instruction. He distrusted abstractions that could not be directly related to the facts. When, several years later, I wrote a general-interest article on the Connection Machine for *Scientific American*, he was disappointed that it left out too many details. He asked, "How is anyone supposed to know that this isn't just a bunch of crap?"

Feynman's insistence on looking at the details helped us discover the potential of the machine for numerical computing and physical simulation. We had thought that the Connection Machine would not be efficient at "number crunching," because the first prototype had no special hardware for vectors or floating-point arithmetic. Both of these were "known" to be requirements for number crunching. Feynman decided to test this assumption on a problem he was familiar with in detail: quantum chromodynamics.

Quantum chromodynamics is the presently accepted field theory of the strongly interacting elementary particles in terms of their constitutent quarks and gluons. It can, in principle, be used to compute the mass of the proton (in units of the pion mass). In practice, such a computation might require so much arithmetic that it would keep the fastest computers in the world busy for years. One way to do the calculation is to use a discrete four-dimensional lattice to model a section of space-time. Finding the solution involves adding up the contributions of all the possible configurations of certain matrices at the links of the lattice, or at least some large representative sample. (This is essentially a Feynman path integral.) What makes this so difficult is that calculating the contribution of even a single configuration involves multiplying the matrices around every loop in the lattice, and the number of loops grows as the fourth power of the lattice size. Because all these multiplications can take place concurrently, there is plenty of opportunity to keep all 64 000 processors busy.

To find out how well this would work in practice, Feynman had to write a computer program for quantum chromodynamics. Because BASIC was the only computer language Richard was really familiar with, he made up a parallel-processing version of BASIC in which he wrote the program. He then simulated the operation of the program by hand to estimate how fast it would run on the Connection Machine.

He was excited by the results. "Hey Danny, you're not gonna believe this, but that machine of yours can actually do something *useful!*" According to Feynman's calculations, the Connection Machine, even without any special hardware for floating-point arithmetic, would outperform a machine that Caltech was building explicitly for quantum chromodynamics calculations. From that point on, Richard pushed us more and more toward looking at numerical applications of the machine.

By the end of that summer of 1983, Richard had completed his analysis of the behavior of the router, and much to our surprise and amusement, he presented his answer in the form of a set of partial differential equations. To a physicist this may seem natural, but to a computer designer it seems a bit strange to treat a set of Boolean circuits as a continuous, differentiable system. Feynman's router equations were written in terms of variables representing continuous quantities such as "the average number of 1 bits in a message address." I was much more accustomed to inductive proof and case analysis than to taking the time derivative of "the number of 1's." Our discrete analysis said we needed seven buffers per chip; Feynman's differential equations suggested we only needed five. We decided to play it safe and ignore Feynman.

The decision to ignore Feynman's analysis was made in September, but by the following spring we were up against a wall. The chips we had designed were slightly too big to manufacture, and the only way to solve the problem was to cut the number of buffers per chip back to five. Because Feynman's equations claimed we could do this safely, his unconventional methods of analysis started looking better and better to us. We decided to go ahead and make the chips with the smaller number of buffers.

Fortunately, Feynman was right. When we put together the chips, the machine worked. The first program run on the machine was John Horton Conway's Game of Life, in April 1985.

Cellular Automata

The Game of Life is an example of a class of computations that interested Feynman: cellular automata. Like many physicists who had spent their lives going to successively lower levels of subatomic detail, Feynman often wondered what was at the bottom. One possible answer was a cellular automaton. The notion is that the space-time continuum might ultimately be discrete, and that the observed laws of physics might simply be large-scale consequences of the average behavior of tiny cells. Each cell could be a simple automaton that obeys a small set of rules and communicates only with its nearest neighbors—like the points in the lattice calculation for quantum chromodynamics. If the universe in fact works this way, there should be testable consequences, such as an upper limit on the density of information per cubic meter of space.

The notion of cellular automata goes back to John von Neumann and Stanislaw Ulam, whom Feynman had known at Los Alamos. Richard's recent interest in the subject was aroused by his friends Ed Fredkin and Stephen Wolfram, both of whom were fascinated by cellular automata as models of physics. Feynman was always quick to point out to them that he considered their specific models "kooky," but like the Connection Machine, he considered the subject crazy enough to put some energy into.

There are many potential problems with cellular automata as a model of physical space and time—for example, finding a set of rules that gives relativistic invariance at the observable scale. One of the first problems is just making the physics rotationally invariant. The most obvious patterns of cellular automata, such as a fixed three-dimensional grid, have preferred directions along the grid axes. Is it possible to implement even Newtonian physics on a fixed lattice of automata?

Feynman had a proposed solution to the anisotropy problem that he attempted (without success) to work out in detail. His notion was that the underlying automata, rather than being connected in a regular lattice like a grid or a pattern of hexagons, might be randomly connected. Waves propagating through this medium would, on average, propagate at the same rate in every direction.

Cellular automata started getting attention at Thinking Machines in 1984 when Wolfram suggested that we should use such automata not as a model of nature, but as a practical approximation method for simulating physical systems. Specifically, we could use one processor to simulate each cell with neighbor-interaction rules chosen to model something useful, like fluid dynamics. Wolfram was at the Institute for Advanced Study in Princeton, but he was also spending time at Thinking Machines.

For two-dimensional problems there was a neat solution to the anisotropy problem. It had recently been shown that a hexagonal lattice with a simple set of rules gives rise to isotropic behavior on the macroscopic scale. Wolfram did a simulation of this kind with hexagonal cells on the Connection Machine. It produced a beautiful movie of turbulent fluid flow in two dimensions. Watching the movie got all of us, especially Feynman, excited about physical simulation. We all started planning additions to the hardware, such as support for floating-point arithmetic, which would make it possible to perform and display a variety of simulations in real time.

Feynman the Explainer

In the meantime, we were having a lot of trouble explaining to people what we were doing with cellular automata. Eyes tended to glaze over when we started talking about state transition diagrams and finite-state machines. Finally Feynman told us to explain it like this:

> We have noticed in nature that the behavior of a fluid
> depends very little on the nature of the individual particles
> in that fluid. For example, the flow of sand is very similar

to the flow of water or the flow of a pile of ball bearings. We have therefore taken advantage of this fact to invent a type of imaginary particle that is especially simple for us to simulate. This particle is a perfect ball bearing that can move at a single speed in one of six directions. The flow of these particles on a large enough scale is very similar to the flow of natural fluids.

This was a typical Feynman explanation. On the one hand, it infuriated the experts who had worked on the problem because it did not even mention all of the clever problems that they had solved. On the other hand, it delighted the listeners because they could walk away with a real understanding of the calculation and how it was connected to physical reality.

We tried to take advantage of Richard's talent for clarity by getting him to criticize the technical presentations we made in our product introductions. Before the commercial announcement of the first Connection Machine, CM-1, and all of our subsequent products, Richard would give a sentence-by-sentence critique of the planned presentation. "Don't say 'reflected acoustic wave.' Say *echo*." Or, "Forget all that 'local minima' stuff. Just say there's a bubble caught in the crystal and you have to shake it out." Nothing made him angrier than making something simple sound complicated.

Getting Richard to give advice like that was sometimes tricky. He pretended not to like working on any problem that was outside his claimed area of expertise. Often, when one of us asked for him advice, he would gruffly refuse with, "That's not my department." I could never figure out just what his department was, but it didn't matter anyway, because he spent most of his time working on these "not my department" problems. Sometimes he really would give up, but more often than not he would come back a few days after his refusal and remark, "I've been thinking about what you asked the other day and it seems to me... ." This worked best if you were careful not to expect it.

I do not mean to imply that Richard was hesitant to do the "dirty work." In fact he was always volunteering for it. Many a visitor at Thinking Machines was shocked to see that we had a Nobel laureate soldering circuit boards or painting walls. But what Richard hated, or at least pretended to hate, was being asked to give advice. So why were people always asking him for it? Because even when Richard didn't understand, he always seemed to understand better than the rest of us. And whatever he understood, he could make others understand as well. Richard made people feel like children do when a grown-up first treats them as adults. He was never afraid to tell the truth, and however foolish your question was, he never made you feel like a fool.

The charming side of Richard helped people forgive him for his less charming characteristics. For example, in many ways Richard was a sexist. When it came time for his daily bowl of soup, he would look around for the nearest "girl" and ask if she would bring it to him. It did not matter if she was the cook, an engineer or the president of the company. I once asked a female engineer who had just been a victim of this treatment if it bothered her. "Yes, it really annoys me," she said. "On the other hand, he's the only one who ever explained quantum mechanics to me as if I could understand it." That was the essence of Richard's charm.

A Kind of Game

Richard worked at the company on and off for the next five years. Floating-point hardware was eventually added to the machine, and as the machine and its successors went into commercial production, they were being used more and more for the kind of numerical simulation problems Richard had pioneered with his quantum chromodynamics program. Richard's interest shifted from the construction of the machine to its applications. As it turned out, building a big computer is a good excuse for talking with people who are working on some of the most exciting problems in science. We started working with physicists, astronomers, geologists, biologists, chemists—each of them trying to solve some problem that couldn't have been solved before. Figuring out how to do such calculations on a parallel machine required understanding their details, which was exactly the kind of thing Richard loved to do.

For Richard, figuring out these problems was a kind of game. He always started by asking very basic questions like, "What is the simplest example?" or "How can you tell if the answer is right?" (He asked questions until he had reduced the problem to some essential puzzle he thought he could solve.) Then he would set to work, scribbling on a pad of paper and staring at the results. While he was in the middle of this kind of puzzle-solving, he was impossible to interrupt. "Don't bug me. I'm busy," he would say without even looking up. Eventually he would either decide the problem was too hard (in which case he lost interest), or he would find a solution (in which case he spent the next day or two explaining it to anyone who would listen). In this way he helped work on problems in database searching, geophysical modeling, protein folding, image analyzing and the reading of insurance forms.

The last project I worked on with Richard was in simulated evolution. I had written a program that simulated the evolution of populations of sexually reproducing creatures over hundreds of thousands of generations. The results were surprising, in that the fitness of the population made progress in sudden leaps rather than by the expected steady improvement. The fossil record shows some evidence that real biological evolution might also exhibit such "punctuated equilibrium," so Richard and I decided to look more closely at why it was happening. He was feeling ill by that time, so I went out and spent the week with him in Pasadena. We worked out a model of evolution of finite populations based on the Fokker-Planck equations. When I got back to Boston, I went to the library and discovered a book by Motoo Kimura on the subject. Much to my disappointment, all our "discoveries" were covered in the first few pages. When I called Richard and told him what I had found, he was elated. "Hey, we got it right!" he said. "Not bad for amateurs."

In retrospect I realize that in almost everything we worked on together, we were both amateurs. In digital physics, neural networks, even parallel computing, we never really knew what we were doing. But the things that we studied were so new that none of the others working in these fields knew exactly what they were doing either. It was amateurs who made the progress.

Telling the Good Stuff You Know

Actually, I doubt that it was "progress" that most interested Richard. He was always searching for patterns, for connections, for a new way of looking at something, but I suspect his motivation was not so much to understand the world, as it was to find new ideas to explain. The act of discovery was not complete for him until he had taught it to someone else.

I remember a conversation we had a year or so before his death, walking in the hills above Pasadena. We were exploring an unfamiliar trail, and Richard, recovering from a major operation for his cancer, was walking more slowly than usual. He was telling a long and funny story about how he had been reading up on his disease and surprising his doctors by predicting their diagnosis and his chances of survival. I was hearing for the first time how far his cancer had progressed, so the jokes did not seem so funny. He must have noticed my mood, because he suddenly stopped the story and asked, "Hey, what's the matter?"

I hesitated. "I'm sad because you're going to die."

"Yeah," he sighed, "that bugs me sometimes too. But not so much as you think." And after a few more steps, "When you get as old as I am, you start to realize that you've told most of the good stuff you know to other people anyway."

We walked along in silence for a few minutes. Then we came to a place where another trail crossed ours and Richard stopped to look around at the surroundings. Suddenly a grin lit up his face. "Hey," he said, all trace of sadness forgotten, "I bet I can show you a better way home."

And so he did.

This article was originally published in the February 1989 issue of *Physics Today*.

RICHARD FEYNMAN AT LA CAÑADA HIGH SCHOOL: FEYNMAN'S LAST PUBLIC PERFORMANCE

John S. Rigden

The only time I talked with Richard Feynman was on November 14, 1987—three months plus one day before he died. On that occasion, he was obviously ill, yet his spirit animated the high school auditorium where our panel sat.

For me, it started in late September or early October. I was scheduled to give a colloquium at California Polytechnic State University, San Luis Obispo on Thursday, November 12. Sometime around October 1, I got a call from a member of the physics department at San Luis Obispo. I shall never forget the question posed to me: would I be willing to stay an extra couple of days, go to a meeting in the Los Angeles area, serve on a panel at the meeting, and "fill in for Feynman!" I laughed and said, "Right. I'll fill in for Feynman. You've got to be kidding." The explanation followed: Feynman had agreed to serve on this panel, but he was too ill and had to withdraw. In typical Feynman fashion, he did not want his name to appear on the program so only a few people knew of his potential participation. When I learned that the subject to be discussed by the panel was, "What High School Physics

Should Include," I agreed to be a panel member. Besides, I thought, someday I can tell my grandchildren: I once filled in for Richard Feynman.

On November 12 I was in San Luis Obispo. Sometime during that day, a telephone call came to the physics department informing my hosts that Feynman was feeling better and would participate in the meeting on Saturday (November 14). On hearing this, I immediately offered to withdraw from the panel and happily listen to Feynman and the other panel members from the audience. "No," I was told, "your name is on the program. We shall simply add another chair."

Two days later, on November 14, I stood in the foyer of the auditorium of La Cañada High School in La Cañada, California. About 30 minutes before the meeting was to begin, I saw David Goodstein, a Cal Tech physicist, and Richard Feynman coming up the sidewalk and they entered the foyer. (As I looked at Feynman I was shocked. I had attended a lecture Feynman gave in 1983 and the change in his appearance from that earlier time was jolting.) David saw me and, since we know each other, he walked to me with Feynman at his side. "Dick Feynman," David said, "this is John Rigden." And with that, David walked away.

I remember the thoughts that raced through my mind as I stood there alone with Feynman: "What do I say to this guy? What do I call him? I can't call him Dick, I won't call him Professor Feynman." In a nervous way I blurted out,

"Feynman, what are you doing here? I'm taking your place."

"Ohhhh," said Feynman, "you're taking my place. Then I'll leave."

"No, no, Feynman," I said quickly, still coping with my nervousness, "You stay. You might say something useful." At that, we both laughed. That was how my only conversation with Richard Feynman began.

My friends who knew Feynman tell me that my irreverent remarks were probably a good way to start our conversation. They may be correct. In any event, Feynman and I had a free flowing discussion for the next 30 minutes. At one point I said,

"That was a nice letter you wrote to David Mermin."

Feynman looked at me and said, "Who's Mermin?"

Who's David Mermin? A little background explanation is needed. Mermin is a physicist at Cornell University. But that's not so important. The significant fact is that Mermin wrote a paper entitled "Bringing Home the Atomic World: Quantum Mysteries for Anybody." This paper was published in the *American Journal of Physics* in October 1981 during my tenure as editor. It was a wonderful paper. With only arithmetic used in his argument, Mermin conceived of a simple device that enabled him to analyze the data from a Bohm-Einstein-Podolsky-Rosen experiment in light of Bell's theorem. The data from Mermin's device is so direct and his analysis was developed with such clarity that Richard Feynman—a peerless expositor—was moved to write a letter on March 30, 1984 to Mermin and say, "One of the most beautiful papers in physics that I know of is yours in the *American Journal of Physics*." Of course, David Mermin is enormously proud of this accolade, but that is another story.

So here I was, talking to Feynman in the foyer of the La Cañada High School auditorium, and Feynman had just asked, "Who's Mermin?"

"You remember," I responded, "he wrote that article in the *American Journal of Physics*."

"I don't remember," Feynman asserted.

"Yes you do," I continued. "The article was about Bell's theorem in which he used these little boxes with three settings... ."

"Oh yes," Feynman interrupted, "oh yes, that was a beautiful paper. He did with three variables what I could only do with six variables. Yes, yes, that was a beautiful paper..." And on he went for another several minutes talking about Mermin's simple handling of a complex topic. Feynman didn't remember the name Mermin, but he remembered the physics which, he said, he had used many times "with proper attribution, of course."

The main part of the program was a panel discussion on the question, "What Should High School Physics Include?" Many interesting questions were posed to the panelists. Feynman's responses revealed his profound commitment to the teaching of physics and the values, as well as the priorities, he brought to his teaching. One discussion topic concerned a regulation under consideration at that time by the State of California—a regulation that would mandate a course in physics for all students. Feynman's response, taken from a transcript of the video tape that was made of this session, took issue with such an approach.

"...this idea of the State making a law that this [physics] should be taught for all students without looking around to see how many teachers they have - who can teach it - will produce a catastrophe. The catastrophe will be to put...a large number of teachers [in the classroom] who do not understand the subject and do not love the subject. This will produce the kind of course without any feel for the subject whatsoever... . The result will be a complete degeneration of the physics. So I think anytime you try to teach the subject without teachers who love the subject, it is doomed to failure and is a foolish thing to do."

"I was on the [mathematics] curriculum committee some years ago and the State had to look at everything that anybody presents—it's kind of a democratic law. And so all kinds of little plans for how to teach elementary arithmetic were sent in and they were all wonderful. One used matchsticks, another teaches base 2, another makes little crossword puzzles with numbers. And the wonderful thing was that every one of these methods were successful. In every case there was evidence of this—they tried it in a class and it worked well. The only trouble is, we are not sure if it would work in a class when we don't know if the ideas are communicated to someone whose expertise is not in this area, or they hadn't invented the idea or had no enthusiasm for it. It was always the one who invented it—who loved the subject and had special students or even ordinary students but who had a special attitude and was going to try a new way to teach it. There was a certain enthusiasm and a special relationship between the teacher and the students which was a kind of excitement, and unless that excitement exists between the teacher and student then I don't think education is worthwhile and it's better not to try education under these circumstances."

Another question posed to the panel was, "How are we going to get more people to take physics?" Feynman began his response by saying that students should not be *required* to take physics.

If they are required, "...not only will we have a lot of physics teachers who don't know physics, but we will have... students in their classrooms who are not interested in the subject and when you have a very large number of students who are not interested in the subject, the whole flavor of the class disappears."

Feynman went on:

"I want to add one other thing. I hear a good deal about

Richard and Joan Feynman. (Photo: Joan Feynman.)

Richard and Joan Feynman & dog "Buddy"
(Photo: Joan Feynman.)

Richard Feynman age 13, Graduating grammar school. (Photo: Joan Feynman.)

The Feynmans, (l to r) Richard, Melville, Lucille, Joan in front.
(Photo: Joan Feynman.)

Richard Feynman at MIT. (Photo: Joan Feynman.)

Joan and Richard Feynman on the Beach in Far Rockaway
(Photo: Joan Feynman.)

Richard Feynman's lab in his bedroom, circa 1931 or 1932. (Photo: Joan Feynman.)

(Photos courtesy of Michelle Feynman.)

teaching physics, that it's a difficult subject. I think if you look at it a little bit differently it's the same thing. What is physics?" Feynman asked. Then he answered his question: "It's supposed to be a description of the physical world. Now if you think, 'I'm *not* going to be teaching physics, rather I'm going to be telling [students] about the physical world.' Does that give you a different idea? What is the physical world like? Does that mean we...start [the course] and spend 2/3 of our time with falling bodies...?"

Again Feynman asked a question:

"What tells you as much as possible about the physical world? We can make a list of things about the physical world that are...delightful. One of the things about the physical world is...that an object falls in uniform acceleration. Let's look at [this] a little bit...you know with the torques and angular momentum. That's not so delightful as the fact that everything is made out of atoms. [Knowing about atoms] you can understand what evaporation is and freezing and when it evaporates its becomes cooler because the fast ones leave and so forth. And in one picture, [the atomic picture] you get a whole lot of ideas. Maybe if we would think in terms of telling what the physical world is like so that they can understand it better."

Feynman went on talking about teaching physics:

"...its a way of thinking about teaching: what you're trying to describe is the wonders and the way things actually are and later on you can talk about the falling bodies after a bit. You don't need to know a hell of a lot—just a little [about the] elasticity of colliding molecules—because they have no friction they never stop, they go on forever. Put a few molecules together that they attract each other. [In this view] you have a tremendous amount of understanding of the world. It might be possible to concentrate on those things and it won't be quite as difficult for the students."

Feynman had looked at physics textbooks and drawn the following conclusions:

> "I looked at [textbooks], for instance, I found when I was reading about the atoms that I didn't need the first part of the book, believe it or not, I didn't have to know how falling bodies worked or anything about it and all that stuff was sort of in the way... . I understand that the chemists teach [about atoms] first and take all the juice out of the physics and they get away with it, teaching about atoms first. Maybe we make a mistake trying to teach our subject entirely historically with the first thing that was discovered...the laws of falling bodies and planets."

"What is the essential thing—that one thing—that makes a physics course a physics course? That one thing? If you have to retain that one thing, what would it be?" This was another question posed to the panel. The panelist that answered immediately before Feynman gave his response said that she would teach the law of gravitation. If you have read the *Feynman Lectures*, you can predict Feynman's answer.

> "I would come back to my atoms. Describe how things work in the world of atoms. Something about electrical phenomena, motion of the electrons through the wires, and so on. Understand the things that are happening out there. What it is that they see. It took us a little while to understand these things. It's not absolutely obvious that everything is made of atoms. But, it's in every newspaper article, it helps explain biology, its the key to chemical phenomena and so on. It has a tremendous width.

> "Now if you wanted to teach the *beauty* of physics, then the law of gravitation has a kind of elegance. It has a mathematical aspect and so on. It depends on your purpose. But as I understand the discussion, you're going to teach a lot of people who are not going into physics and we teach them about the physical world and one of the more interesting things about the physical world is to help them understand it, the atomic way things work, solids and liquids. What's going on when we see things. Also some elec-

trical phenomena. It's kind of mysterious how a television works."

When the panel discussion ended, people from the audience swarmed to the stage of the auditorium and surrounded Feynman. They started asking Feynman questions, physics questions. As I watched, I realized I was witnessing something extraordinary. Feynman's energies grew as he responded to question after question. The outside corners of his eyes were creased by the smiles that played over his face as he talked about physics. His hands and arms cut through the air with increasing vigor as their motions served to complement, even demonstrate, *his* explanations.

"I have a question," a man said and, as he positioned himself in front of Feynman. He held a long copper tube in his left hand and two metal cylinders in his right hand.

"All right," said Feynman, "I'll answer, but if there's a trick, I might miss it."

"No trick," said the physicist and with that he released one of the metal cylinders and it fell rapidly through the copper tube and onto the floor. Then he released the second metal cylinder: it fell s*l*o*w*l*y through the tube and the questioner dramatically moved his right hand to the bottom of the copper tube and caught the cylinder as it emerged.

"It's a magnet," said Feynman.

"That's right," said the physicist holding the tube, "but that's not the question. Suppose the tube were a superconductor. With no i^2R losses, how do you account for the energetics of the falling magnet?" It was a great question, Feynman was challenged, and his virtuoso performance continued.

I was not surprised by Feynman's deftness as I watched him—his reputation in such impromptu situations was well known. It was the enjoyment he exuded as he stood there talking physics with an eager, receptive group of physics teachers that moved me. It was an enjoyment I could feel. When the session ended and Feynman, along with David Goodstein, walked out of the La Cañada High School auditorium, I had the feeling that I was standing on holy ground.

Earlier, in the foyer before the panel session began, I had told Feynman that I thought he had cheated the public in his book *Surely You're Joking*.

"Who are you," said Feynman in response, "to tell an author that you don't like his book and that he cheated the reader?"

"If there is one thing about your life that holds it together," I said, "it is your love of physics and that doesn't come through your book."

Immediately, Feynman shot back, "That was deliberate."

"That's how you cheated the public," I said.

After a just-discernable pause, Feynman said, "That's the next book."

When I heard of Feynman's death, early in the morning on February 16, I experienced an eerie sense of personal loss and, with sadness, I thought of the "next book," a book that Feynman would never write.

The panel session at La Cañada High School was Feynman's last public appearance. During the months prior to this meeting, Feynman experienced a succession of physical heights and depths which took him from feeling relatively good to feeling very bad. He had canceled his agreement to appear on the panel during one of his bad times, but a few days before the meeting he was once again feeling well enough to reaffirm his commitment. So he had his secretary, Helen Tuck, call the organizer of the panel session to indicate his willingness to participate. In spite of his deteriorating health, high school physics was a subject Feynman would *choose* to discuss and La Cañada High School was a place Feynman would *choose* to come. It is somehow fitting that Feynman's last public appearance was in a high school auditorium discussing what should be taught in high school physics.

Feynman-The Man

A LOWBROW'S VIEW OF FEYNMAN

Valentine L. Telegdi

I learned by accident that PHYSICS TODAY was preparing a special issue in memory of Richard Feynman, whose death about a year ago shocked his fellow physicists worldwide so deeply. In looking at the list of contributors to the memorial issue, I noticed with surprise that it consisted almost entirely of theoretical physicists. While eminent theorists obviously were best qualified to eulogize Feynman's extraordinary contributions to their own specialty, something essential might have been lost to the readers by this restricted choice. Feynman was not a theorist's theorist, but a physicist's physicist and a teacher's teacher.

What did Feynman do for the experimental physicist? In my experience, two things. First, he gave us (or gave us back?) our dignity. He, who respected no authority, enabled us to respect ourselves. This he did in two ways: He emphasized that physics was an experimental science, and hence would dry up without facts. He understood experiments deeply and could suggest sources of error that had escaped the experimenters themselves. Being questioned by him was a harrowing but deeply rewarding experience. Beyond that, he gave us the assur-

Valentine L. Telegdi is a Professor Emeritus of Physics at the Swiss Federal Institute of Technology, Zurich, Switzerland and is currently visiting the California Institute of Technology, Pasadena, and CERN, Geneva. He was formerly the Enrico Fermi Distinguished Service Professor at the University of Chicago.

ance, by ripping away the mask of formalism from the face of simple facts, that we too could understand the deeper meaning and significance of our own work. To him, as to Enrico Fermi, the difference between a theorist and an experimenter was one of technique and not primarily one of intellectual competence.

Second, his insight often created experimental opportunities in fields that were barren before. Where would, say, high-energy neutrino physics (a subject of obvious topical interest) be without the parton model? I remember times when only elastic scattering from protons was respectable; Feynman's picture legitimized the scattering from "junk." Similar remarks apply to the study of the hadronic "garbage" that recoils in the scattering act—now a rich source of information as jets.

Had Feynman not reformulated quantum mechanics, had he not invented his diagrams, had he not—with Murray Gell-Mann—laid the cornerstone of weak-interaction theory, he would still be remembered by generations for *The Feynman Lectures on Physics*, his three-volume introductory course (written with Robert Leighton and Matthew Sands). Maybe—Feynman himself often raised this question—that course did not hit the mark as far as the student audience was concerned. But it certainly did and does a lot for the teachers.

To a few of us Dick Feynman was a revered friend, to many of us a unique source of inspiration, and to almost all of us an irreplaceable teacher. His name will remain a household word among experts and beginning students alike.

This article was originally published in the February 1989 issue of *Physics Today*.

R.P. FEYNMAN: THE BEGINNINGS OF A TEACHER

Joan Feynman

I was Richard Feynman's first student and he was my first teacher. We were brother and sister, the only children in our family. When I was a baby, Richard would bundle me into my carriage and take me over to his friend Bernie's house. There he would prop me up so I could watch the two boys work with the batteries, wires, rheostats, switches, and radio tubes they had collected for their "laboratory." He was nine.

I soon graduated to larger tasks. We had a dog, a fox terrier (more or less), the kind you could see in circuses back then. The family taught him tricks, like sitting and begging, by patiently getting him to understand what was expected and then giving him a treat when he was successful. The dog worked hard for the dog biscuits and amazed the neighborhood children. Observing this, Richard decided that I was probably trainable too, and the most amazing trick he could think of to teach me was to do arithmetic. The problem was, what to give me for a treat? Our mother was very careful with our diet and I certainly couldn't have candy between meals. But he was always resourceful.

Joan Feynman is a Physicist on the staff of the Jet Propulsion Laboratory in Pasadena, California.

When I got a problem correct I was allowed to pull his hair until it hurt or, to be more exact, until he grimaced as if in pain. I remember standing in my crib, maybe three years old, yanking on his hair with great delight while he excitedly planned to surprise Bernie with my new trick. I had just learned to add two and three. I have always believed that the reason Richard had a full head of hair all his life was because I had done such a good job of strengthening the roots.

My career continued to develop. At five I was hired as a lab assistant for two cents a week. Richard had an electronics lab in his room and there were several boxes distributed around the floor. He would get everything ready for his next experiment and I would climb up on the boxes so I could reach the switches which I threw on command. My other duty was to put my finger into a small spark gap—again for the edification of his friends. I got quite used to being shocked.

Our world was full of physics. I learned to shuffle my feet on the rug in the long narrow hall in our apartment and then discharge the static electricity by touching the lamp. When we washed the dishes we watched the soap suds trace out the streamlines on the top surface of the water. Spoons dragged through the tea in the cup produced beautiful vortices on each side. At Passover the wine glasses were filled with different amounts of wine and they'd each make different notes when we ran our fingers around the rims, much to our grandmother's consternation. Richard showed me the stars at night from the roof of the apartment house we lived in. Of course, this was only done when the weather was pleasant in the summer, and only early in the evening. As a result the single constellation I know to this day is Orion. When we passed fence posts with a chain swinging between them Richard would whack the chain sharply and we would watch the wave go down the chain and be reflected at the post. He showed me resonances by using trees. If you jar a tree trunk the entire tree shakes a little and each branch moves with its own frequency. Richard would watch and identify the frequency of a selected branch and then push against the tree in the right rhythm. With very little effort he would get large amplitude oscillations in a single branch while the rest of the tree seemed undisturbed. Years later I used this method to knock a rotten branch from high in a tree when I needed some firewood on a camping trip.

My brother's most dramatic demonstration concerned centrifugal force. He used to baby-sit for our parents when they went out for the evening. I was not allowed to get out of bed after my bedtime, so I would lie in bed and call out that I was thirsty. He would come in with a glass of water, swinging it around rapidly in a circle in such a way that the glass was upside down during part of the arc. The trick was well developed; the velocity was always high enough so the water did not come out;—except, of course, the night the glass slipped out of his hand and went soaring across the room.

Besides everyday nature there were a few spectacular events. I remember being about four or five and seeing my first aurora. But the great event of my early childhood was the solar eclipse. I was not yet of kindergarten age when the family excitement began to grow. Preparations were made long before the event. I was told over and over again that I should not look at the sun because it might blind me. Richard was busy exposing pieces of photographic film and stacking them to make filters for us to look through. On the day of the eclipse, my mother hurried home from grocery shopping with me so we would be there on time. Then the moon began to cover the face of the sun and the shadows of the tree leaves on the sidewalk began to take strange shapes because each spot in the dappled shade under trees is a pinhole camera picture of the solar disk. It was terribly eerie. I was being very careful not to look at the sun itself and so be blinded, when Richard arrived with his filters and told me to look at the sun through them. But I was very little and no one had explained that I could safely look through the filters. I was sure I would be blinded. I remember running away as fast as I could with Richard chasing me, waving his filters. I don't know if that eclipse was partial or total. I only know that I never saw it. And to this day I am uncomfortable looking through a telescope at the sun, even though I tell myself over and over that the solar disk is occulted.

The science Richard told me about was not just explanations of the obvious phenomena I could see around me. There were also atoms and molecules, but I must have been seven before I could get it through my head that, not only was the living room chair made of these tiny bits, but I was also. That was a particularly difficult and uncomfortable concept because the atoms were not living and could not think, and I was somehow essentially different, or so it seemed to me. I had more fun with the idea of dinosaurs, so big they could look into the second story window while standing in the street below; or the idea of material from stars, so dense that it would fall right through the floor of the apartment.

The things I have described here about our home may give the impression of a serious learning environment, or some such thing. Nothing could be further from the case. It is true that our father believed that science was the highest calling. Our mother, however, had different ideas. She felt that life was hard and, therefore, it was important to contribute something to make it more endurable. She believed that being a comedian was the highest calling. Under her influence, there was always laughter in our house. There were strict rules about the joking, however. It was never to be done at the expense of anyone else. A cruel comment would surely have met with cold disapproval from our mother, but I never heard a cruel joke at home. The source of our comedy was the absurdity of life. I have wonderful memories of evenings at the supper table when Richard was home from college and he and Mother would get going. My father and I would laugh so hard that our stomachs hurt and we would beg for mercy, but they wouldn't stop until I had fallen off my chair and was literally rolling on the floor.

But nothing is ideal, and for me, the problem was that my interest in science was not taken seriously by my parents. In their world, men became scientists, women became wives and mothers, and somehow the two things were not compatible. When I was eight, my mother even told me that women's brains were incapable of doing science. She meant no harm in telling me this. She believed it to be the objective truth. But I remember curling up in a chair in the living room and crying. This idea had some real consequences for my education. It was decided very early that I would not be sent to a technical university because the men at those schools were believed not to like the women who went there. Moreover, my parents did not encourage me to become active in science. When I was in my early teens and Richard was in graduate school at Princeton, he wanted to buy a piece of optical glass for me to grind a telescope lens. He intended to pay for it out of his own money, and he had very little money at the time. He planned to take it to Princeton to check the grinding. I was delighted. The only thing that was required was that I have a place in which I could grind it. My room was ruled out but we lived in an apartment house in which each of the renters had a small roomlike storage space in the basement. One of the boys in the building had a chemistry lab in one of these, and I often went down there to watch him do experiments. Richard and I suggested I set up my space in our storage room, but this was refused; my mother imagined a world full of men lurking in dark corners of the basement waiting to rape me. And so, since my parents didn't think it was important for their daughter to grind a telescope lens, I never did.

Richard however was of a more independent mind. Although lens grinding fell through, we had a mathematics notebook that we sent back and forth when he was in graduate school. He would write problems in it for me to work on, and I would send my solutions back and also ask any question I wanted about science. Although he hated writing letters, he would always answer my questions.

Then, on my fourteenth birthday, Richard gave me a present that changed my life. He had been visiting from Princeton and when he went back he left a book on the table next to the front door. I took it to my mother, saying that Richard had forgotten a book. But on the way to the kitchen I opened it and saw that he had written my name in it! It was the 1933 edition of *Astronomy*, by R. H. Baker. It was a college text and I was overwhelmed with the idea that I could read it. I would read until I got stuck and then start again. Each time I got a little further. I have it here beside me now as I write, open to page 407 and the figure that has been so important in my life. It's a graph and the caption reads "Relative strengths of the Mg+ absorption line at 4482 angstroms from *Stellar Atmospheres* by Celillia Payne-Gaposhkin."

Payne-Gaposhkin! A *married woman!* A famous astronomer, writing a book that is quoted in a text! The secret was out: It was possible! From that day on, I was able to take my own interest in science seriously.

Richard went away to Los Alamos some time after he gave me the book and our relationship was reduced to correspondence only. Since he hated to write, there were not as many letters as the family could have wanted. We hatched all sorts of schemes to make him write, the most successful of which was to send the letters in code. My father and I would make up a code, write the message and include the key in the envelope, with a note asking the censor to take the code out. Since it was illegal to censor mail within the United States, the existence of the censor was secret information which we were not supposed to know. But the censor cooperated and took the keys out anyway. The most successful code I devised contained symbols without meaning, a kind of noise. That kept him going for awhile.

When I went away to Oberlin College, my father told me that he wanted me to be able to make a living when I graduated. I decided to major in physics with a sort of vague idea that I could get a bachelors degree and get a job helping a real physicist somehow. The lesson of Payne-Gaposhkin had not yet been completely learned. I have a letter that Richard wrote me during that period answering a letter of mine in which I had asked his advice on two matters. The first problem was that I had realized that I was not in love with the man who was planning to marry me, and I didn't know how to tell him. Richard's advice was to tell him quickly and clearly. The second problem was that I didn't know what I ought to do about having a career. I had written that I thought perhaps I could be a mediocre physicist. I wanted his response to this idea. Did he think it was possible? He gave me the advice which I quote here (emphasis his):

> Now about your career. That is even easier. Do what you *want*. Sometimes that isn't possible, because of financial difficulties, prejudice against women, a certain field looks too difficult, etc. Only please don't *plan* on being mediocre. Then you *certainly* will not be otherwise. But if you plan on being great you work much harder, and even if you turn out mediocre, you will be a lot better medium than if you hadn't tried for an ideal.

Now never in the history of the world did a little sister take a big brother's advice, and I did not take Richard's advice on my career. But I give it here in the hopes that some other little sister will take heed.

After the war, I was married and had my children and career in the East, while Richard went to live in California with his family. We drifted apart somewhat, although I think we were always closer than most brothers and sisters. After many years, my youngest child went away to college and I was living alone in Boston. One February day in 1984, I was looking out the window at the falling snow when the thought came to me "What am I doing here? Where would I rather be?" Richard already had cancer and I realized that if I ever was going to spend more time with him it had better be soon. So I called some friends at the Jet Propulsion Laboratory in Pasadena and told them I wanted to come out. I was lucky and the next fall I joined the lab and rekindled the relationship with Richard.

Things hadn't changed much. He had become older and more famous, but he was just as excited about life and science as he had been as a kid. He liked to laugh just as much, too. Over the years, he had decided always to do as he pleased. This meant doing science and taking on no administrative or government advisory responsibilities. He delighted in what he called his "irresponsibility." Of course, the serious reason for this so-called "irresponsibility" was that he believed that if a person was good at one thing, it didn't necessarily mean he was good at another. He knew he was good at physics, but doubted that he would be good at administration and science management. He was absolutely sure that his talents did not include the ability to advise the government. He believed it was OK to do what he liked and was good at.

All his life, he had done physics for fun and he was still doing it for fun. He said that when people asked him how long he worked each week, he really couldn't say, because he never knew when he was working and when he was playing.

The years I spent with him in Pasadena were good in spite of his illness. Both he and his wife Gweneth had cancers that would prove fatal. However, both of them were always too busy living to be stopped by their illnesses. They had all sorts of adventures during those years, together and separately. They went to little villages off the beaten track in Japan. Gweneth planned and built gardens for her family and for friends. Richard wrote amusing books and served on the commission investigating the shuttle disaster. Together they worked on elaborate schemes to get to Tuva, a small far away part of the Soviet Union that they called "the pole of inaccessibility."

Their daily life went on peacefully. I would visit their home on Thursday nights and Gweneth always cooked something special for me, as she did for every guest. Richard sat at the head of the table and would look at Gweneth and the rest of his family with evident pride and contentment. Then, after supper, Richard and I would sit in the living room for many hours, talking. Sometimes on weekends we would take long walks in the canyons near Pasadena or along the beach. We talked about all kinds of subjects, limited only by the requirement that we were both acknowledged nonexperts. We listened to each others ideas carefully, knowing that we were undoubtedly wrong, but enjoying the exercise of inventing new ideas, considering them, turning them upside down and playing with them. Since neither one of us really knew what we were talking about, we talked as equals, equally ignorant, of course, but equals. We talked about the evolution of the mind, hypnosis, human nature, war, the scientific process, why the sea was boiling hot, and whether pigs had wings.

My first teacher was undoubtedly the best. But I prefer to remember him as simply my brother, sharing his wonder and companionship over the many years of both our lives.

[Biographical information: Joan Feynman received her B. A. in physics from Oberlin College and her Ph.D. in theoretical solid state physics from Syracuse University. She is the mother of three grown children. In 1963 she became interested in space physics and has been working in that field ever since. She is author and coauthor of over 70 papers. She has served as associate editor of the *Journal of Geophysical Research,* the leading journal in her field, and was an officer of her professional organization, the American Geophysical Union. She is now a member of the technical staff at the Jet Propulsion Laboratory, California Institute of Technology, Pasadena, CA.]

BIBLIOGRAPHY OF RICHARD P. FEYNMAN *

Prepared by N. Anderson, J. Goodstein, B. Ludt: Cal Tech Archives

1939

With M. S. Vallarta, Scattering of cosmic rays by the stars of a galaxy, Phys. Rev., **55**, 506–507.

Forces in molecules, Phys. Rev., **56**: 340–343.

1942

The Principle of Least Action in Quantum Mechanics (thesis). Princeton. Ann Arbor: University Microfilms.

1945

With J. A. Wheeler. Interaction with the absorber as the mechanism of radiation. Rev. Mod. Phys. **17**: 157–181.

1946

Amplifier response. Edited and declassified work from the Manhattan Project. 15 pgs.

1948

Space time approach to non-relativistic quantum mechanics. Rev. Mod. Phys., **20**: 367–387.

A Relativistic cut-off for classical electrodynamics. Phys. Rev., **74**: 939–946.

Relativistic cut-off for quantum electrodynamics. Phys. Rev., **74**: 1430–1438.

1949

With J. A. Wheeler. Classical electrodynamics in terms of direct inter-particle action. Rev. Mod. Phys. **21**: 425–433.

With N. Metropolis and E. Teller. Equations of state of elements based on the generalized Fermi-Thomas theory. Phys. Rev., **75**: 1561–1573.

The theory of positions. Phys. Rev. **76**: 749–759.

Space-time approach to quantum electrodynamics. Phys. Rev., **76**: 769–789.

1950

Mathematical formulation of the quantum theory of electromagnetic interaction. Phys. Rev., **80**: 440–457.

1951

An operator calculus having application in quantum electrodynamics. Phys. Rev., **84**: 108–128.

With Carl W. Helstrom, Malvin A. Ruderman and William Karzas. *High Energy Phenomena and Meson Theories*: Notes on course at CIT January-March 1951.

The concept of probability in quantum mechanics (Second Berkeley symposium on mathematical statistics and probability, 1950, University of California, Berkeley), 553–541.

1952

With L. M. Brown. Radiative correction to Compton scattering. Phys. Rev., **85**: 231–244.

1953

The lambda transition in liquid helium. Phys. Rev., **90**: 1116–1117.

Atomic theory of the lambda transition in helium. Phys. Rev., **91**: 1291–1301.

Atomic theory of liquid helium near absolute zero. Phys. Rev., **91**: 1301–1308.

Atomic theory of liquid helium. Reprint from the proceedings of the International Conference on Theoretical Physics, Kyoto and Tokyo, Japan. 895–901.

With M. Baranger and H. A. Bethe. Relativistic correction to the Lamb shift. Phys. Rev., **92**: 482–501.

1954

Atomic theory of the two-fluid model of liquid helium. Phys. Rev., **94**: 262–277.

With G. Speisman. Proton-neutron mass difference. Phys. Rev., **94**: 500.

The present situation in fundamental theoretical physics.

Academia Basileira de Ciencias, **26**: 51–59.

1955

Application of quantum mechanics to liquid helium. In: *Progress in Low Temperature Physics*, Vol. 1. Amsterdam: North Holland, pp. 17–53.

Slow electrons in a polar crystal. Phys. Rev., **97**: 660–665.

With M. Cohen. The character of the roton state in liquid helium. Prog. of Theoretical Phys., **14**: 261–262.

The value of science. Engineering and Science, **19**: 13–15 (Dec.).
Helium II in rotational flow. Science, **121**: 622.

1956

With M. Cohen. Energy spectrum of the excitations in liquid helium. Phys. Rev., **102**: 1189–1204.

With F. deHoffmann and R. Serber. Dispersion of the neutron emission in U-235 fission. High Energy, **3**: 64–69.

The relation of science and religion. Engineering and Science, **19**: 20–23 (June).

1957

Superfluidity and superconductivity. Rev. of Mod. Phys., **29**: 205–212.

The role of science in the World Today. Proceedings of the Institute of World Affairs, **33**: 17–31.

With M. Cohen. Theory of inelastic scattering of cold neutrons from liquid helium. Phys. Rev., **107**: 13–24.

With F. L. Vernon, Jr. and R. W. Hellwarth. Geometrical representation of the Schrodinger equation for solving maser problems. J. Appl. Phys., **28**: 49–52.

An historic moment in physics. Engineering and Science, **20**: 17–18 (Feb.).

1958

With M. Gell-Mann. Theory of the Fermi interaction. Phys. Rev., **109**: 193–198.

Excitation in liquid helium. Physics 24, Kamerlingh Onnes Conference, Leiden. S18–26.

1959

Series of Lectures on the Theory of Fundamental Processes. Pasadena: CIT.

1960

There's plenty of room at the bottom. Engineering and Science, **23**: 22–36 (Feb.).

1961

The present status of quantum electrodynamics. Extrait des rapports et discussions, Solvay, Institut International de Physique.

Quantum Electrodynamics; a lecture note and reprint volume. New York: W. A. Benjamin. In: Frontiers in physics. (German edition, Munich, 1989).

The Theory of Fundamental Processes; a lecture note volume. New York: W. A. Benjamin. In: Frontiers in physics.

1962

With R. W. Hellwarth, C. K. Iddings, and R. M. Platzman.
Mobility of slow electrons in a polar crystal. Phys. Rev., 127:1004–1017.

With R. S. Edgar, S. Klein, I. Lielausis, and C. M. Steinberg. Mapping experiments with r mutants of bacteriophage T4D1. Genetics, 47:179–185.

Lectures in Elementary Physics. Pasadena: CIT.

Quantum Electrodynamics: a lecture note and reprint volume. In: Frontiers in physics. New York: W. A. Benjamin. (German edition, Munich, 1989).

Quantum Electrodynamics: a lecture note and reprint volume. 2nd printing with revisions. In: Frontiers in physics. Reading, MA: Benjamin/Cummings. (German edition, Munich, 1989).

The Theory of Fundamental Processes: a lecture note volume. 2nd printing, corrected. In: Frontiers in physics. New York: W. A. Benjamin.

1963

The problem of teaching physics in Latin America. Engineering and Science, 27:21–30 (Nov.).

With F. L. Vernon, Jr. The theory of a general quantum system interacting with a linear dissipative system. Annals of Phys., 24:118–173.

With Robert B. Leighton and Matthew Sands. *The Feynman Lectures on Physics*, Vol. I. Reading, MA: Addison-Wesley. (Italian/English edition, London/Reading, MA, 1968; Polish edition, Warsaw, 1968, 1971, 1974; French/English edition, London/Reading, MA, 1969; Hungarian edition, Budapest, 1969; Romanian edition, Bucurest, 1969; German/English edition, Munich, 1971; Spanish/English edition, Bogota, 1971; Slovak edition, Bratislava, 1980, 1985; German edition, Munich, 1987).

The quantum theory of gravitation. Acta Physica Polonica, 24:697–722.

1964

With Robert B. Leighton and Matthew Sands. *The Feynman Lectures on Physics*, Vol. II. Reading, MA: Addison-Wesley. (Italian/English edition, London/Reading, MA, 1968; Polish edition, Warsaw, 1968, 1971, 1974; French/English edition, London/Reading, MA, 1969; Hungarian edition, Budapest, 1969; Romanian edition, Bucurest, 1969; German/English edition, Munich, 1971; Spanish/English edition, Bogota, 1971; Slovak edition, Bratislava, 1980, 1985; German edition, Munich, 1987).

The quantum theory of gravitation. In: *Proc. on Theory of Gravitation.* Paris: Gauthier-Villars.

The quantum theory of gravitation. In: *Proc. on Theory of Gravitation.* Warsaw: PWN-Polish Scientific Publishers, pp. 207–208.

With M. Gell-Mann and G. Zweig. Group U(6) $\times$ U(6) generated by current components. Phys. Rev. Letters, 13:678–680.

Lectures on Physics: Exercises. Reading, MA: Addison-Wesley.

1965

With Robert B. Leighton and Matthew Sands. *The Feynman Lectures on Physics,* Vol. III. Reading, MA: Addison-Wesley. (Italian/English edition, London/Reading, MA, 1968; Polish edition, Warsaw, 1968, 1971, 1974; French/English edition, London/Reading, MA, 1969; Hungarian edition, Budapest, 1969; Romanian edition, Bucurest, 1969; German/English edition, Munich, 1971; Spanish/English edition, Bogota, 1971; Slovak edition, Bratislava, 1980, 1985; German edition, Munich, 1987).

With Albert R. Hibbs. *Quantum Mechanics and Path Integrals.* In: International series in pure and applied physics. New York: McGraw-Hill.

Consequences of SU, symmetry in weak interactions. In: *Symmetries in Elementary Particle Physics,* "Ettore Majorana," NY/London: Academic Press, pp. 111–174.

New textbooks for the "new" mathematics. Engineering and Science, 28:9–15 (March).

New textbooks for the "new" mathematics. The California Institute of Technology Quarterly, 6:2–9 (Spring).

The Character of Physical Law. Cambridge: MIT Press. In: The Messenger lectures, Cornell University, 1964; The MIT Press paperback series, 66; (French edition, Paris, 1970; Italian edition, Torino, 1971; Spanish edition, Santiago de Chile, 1972; Serbo-Croatian edition, Zagreb, 1977; Hungarian edition, Budapest, 1983; Spanish edition, Barcelona, 1983; Portuguese edition, Lisbon, 1989).

1966

The development of the space-time view of quantum electrodynamics. Science, 153:699–708.

The development of the space-time view of quantum electrodynamics. Physics Today, 19:??.

What is and what should be the role of scientific culture in modern society. Supplemento al Nuovo Cimento, 4:492–524.

The development of the space-time view of quantum electrodynamics. In: *Les Prix Nobel 1965*. Stockholm: Imprimerie Royale P. A. Norstedt & Soner, pp. 172–191.

1967

The Character of Physical Law. Cambridge, MIT Press. In: The Messenger lectures, 1964; The MIT Press paperback series, 66; (French edition, Paris, 1970; Italian edition, Torino, 1971; Spanish edition, Santiago de Chile, 1972; Serbo-Croatian edition, Zagreb, 1977; Hungarian edition, Budapest, 1983; Spanish edition, Barcelona, 1983; Portuguese edition, Lisbon, 1989).

1969

Very high-energy collisions of hadrons. Phys. Rev. Letters, 23:1415–1417.

The behavior of hadron collisions at extreme energies. In: *High Energy Collisions*. London: Gordon and Breach, pp. 237–256.

Present status of strong electromagnetic and weak interactions. Ceskoslovensky Casopis pro Fyziku, A19:47–59.

With R. B. Leighton and R. E. Vogt. *Exercises in Introductory Physics*. Reading, MA: Addison-Wesley.

What is science? The Physics Teacher, 9/69:313–320.

The application of mathematics to mathematics. American Mathematical Monthly, 76:1178–79.

1970

With S. Pakvasa and S. F. Tuan. Some comments on baryonic states. Phys. Rev., D2:1267–1270.

With K. K. Thorber. Velocity acquired by an electron in a finite electric field in a polar crystal. Phys. Rev., B1:4099–4114. Errata. Phys. Rev., B4:674.

1971

With M. Kislinger and F. Ravndal. Current matrix elements from a relativistic quark model. Phys. Rev., D3:2706–2732.

Closed loop and tree diagrams. W. H. Freeman and Co.

Problems in quantizing the gravitational field, and the massless Yang-Mills field. W. H. Freeman and Co.

Lectures on Gravitation. Pasadena: CIT.

1972

Statistical Mechanics: a set of lectures. Reading, MA: W. A. Benjamin. In: Frontiers in physics. (Russian edition, Moscow, 1975; Polish edition, Warsaw, 1980).

Photon-Hadron Interactions. Reading, MA: W. A. Benjamin. In: Frontiers in Physics. (Russian edition, Moscow, 1975).

What neutrinos can tell us about partons. In: *Proc. of Neutrino '72 Europhysics Conf.,* Vol II. Budapest: OMKD Technoinform, pp. 75–96.

Fisica de Atlas Energias: Cursos de Verano 1972. Mexico: COPAA-SEDICT.

1973

Quarks. Fizikai Szemle, 23:1–7.

1974

Take the world from another point of view. Engineering and Science, 37:11–13 (Feb.).

Cargo cult science. Engineering and Science, 37:10–13 (June).

Partons. In: *Proc. of the 5th Hawaii Topical Conference in Particle Physics*. Honolulu: Univ. Press of Hawaii, pp. 1–97.

Structure of the photon. Science, 183:601–610.

1976

Los Alamos from below: reminiscences of 1943–1945. Engineering and Science, 39:11–30 (Jan.-Feb.).

1977

With R. D. Field and Geoffrey Fox. Correlations among particles and jets produced with large transverse momenta. Nucl. Phys., B128:1–65.

With R. D. Field. Quark elastic scattering as a source of high-transverse-momentum mesons. Phys. Rev., D15:2590–2616.

Gauge theories. In: *Weak and Electromagnetic Interactions at High Energy*. Amsterdam/New York: North Holland, pp. 121–204.

1978

With R. D. Field and Geoffrey Fox. A quantum-chromodynamic approach for the large-transverse-momentum production of particles and jets. Phys. Rev., D18:3320–3343.

With R. D. Field. A parametrization of the properties of quark jets. Nucl. Phys., B136:1–76.

1981

The qualitative behavior of Yang-Mills theory in $2+1$ dimensions. Nucl. Phys., B188:479–512.

1982

Simulating physics with computers. Int. J. of Theo. Phys., 21:6–7.

A qualitative discussion of quantum chromodynamics in $2+1$ dimensions. In: *Proc. Int. Conf. on High Energy Physics, Lisbon, July 9–15, 1981*. Geneva: European Physical Society, pp. 660–683.

Partons. In: *Hawaii Topical Conference in Particle Physics*, Vol. I: Selected Lectures. Singapore, World Scientific Publishers, pp. 229–424.

1984

With Ralph Leighton. The dignified professor. Engineering and Science, 48:4–10 (Nov.).

1985

Quantum mechanical computers. Opt. News, 11:11–46.

QED: The Strange Theory of Light and Matter. Princeton: Princeton U. P. In: Alix A. Mautner memorial lectures. (Japanese edition, Tokyo, 1987; Dutch edition, Amsterdam, 1988; Portuguese edition, Lisbon, 1988; Russian edition, Moscow, 1988).

With Ralph Leighton and Edward Hutchings. *Surely You're Joking, Mr. Feynman: adventures of a curious character*. New York: W. W. Norton.

The computing machines in the future. Nishina Memorial Lecture, Nishina Foundation and Gakushuin.

1986

With H. Kleinert. Effective classical partition functions. Phys. Rev., A34:5080–5084.

Quantum mechanical computers. Foundations of Phys., 16:507–531.

1987

With S. Weinberg. *Elementary Particles and the Laws of Physics: the 1986 Dirac memorial lectures*. NY: Cambridge U. P.

Mr. Feynman goes to Washington. Engineering and Science, 51:6–22 (Fall).

Was ist Naturwissenschaft? Physik und Didaktik, 2:105–116.

1988

Difficulties in applying the variational principle to quantum field theories. In: *Proc. of the International Workshop on Variational Calculations in Quantum Field Theory, Wangerooge, West Germany, Sept. 1–4, 1987.* Singapore: World Scientific Press.

With Ralph Leighton. *What Do You Care What Other People Think.* New York: W. W. Norton.

An outsider's inside view of the Challenger inquiry. Phys. Today, 41:26–37.

1989

The Feynman Lectures on Physics. Reading, MA: Addison-Wesley.

* We gratefully acknowledge the help of the archivist, Judy Goodstein, at the California Institute of Technology.

Laurie M. Brown, John S. Rigden